CALIBRATION TECHNIQUES IN NYQUIST A/D CONVERTERS

THE INTERNATIONAL SERIES IN ENGINEERING AND COMPUTER SCIENCE

ANALOG CIRCUITS AND SIGNAL PROCESSING
Consulting Editor*: Mohammed Ismail. *Ohio State University

Related Titles:

WIDE-BANDWIDTH HIGH-DYNAMIC RANGE D/A CONVERTERS
Doris,Konstantinos, van Roermund, Arthur, Leenaerts, Domine
Vol. 871 ISBN: 0-387-30415-0
METHODOLOGY FOR THE DIGITAL CALIBRATION OF ANALOG CIRCUITS AND SYSTEMS: WITH CASE STUDIES
Pastre, Marc, Kayal, Maher
Vol. 870, ISBN: 1-4020-4252-3
HIGH-SPEED PHOTODIODES IN STANDARD CMOS TECHNOLOGY
Radovanovic, Sasa, Annema, Anne-Johan, Nauta, Bram
Vol. 869, ISBN: 0-387-28591-1
LOW-POWER LOW-VOLTAGE SIGMA-DELTA MODULATORS IN NANOMETER CMOS
Yao, Libin, Steyaert, Michiel, Sansen, Willy
Vol. 868, ISBN: 1-4020-4139-X
DESIGN OF VERY HIGH-FREQUENCY MULTIRATE SWITCHED-CAPACITOR CIRCUITS
U, Seng Pan, Martins, Rui Paulo, Epifânio da Franca, José
Vol. 867, ISBN: 0-387-26121-4
DYNAMIC CHARACTERISATION OF ANALOGUE-TO-DIGITAL CONVERTERS
Dallet, Dominique; Machado da Silva, José(Eds.)
Vol. 860, ISBN: 0-387-25902-3
ANALOG DESIGN ESSENTIALS
Sansen, Willy
Vol. 859, ISBN: 0-387-25746-2
DESIGN OF WIRELESS AUTONOMOUS DATALOGGER IC'S
Claes and Sansen
Vol. 854, ISBN: 1-4020-3208-0
MATCHING PROPERTIES OF DEEP SUB-MICRON MOS TRANSISTORS
Croon, Sansen, Maes
Vol. 851, ISBN: 0-387-24314-3
LNA-ESD CO-DESIGN FOR FULLY INTEGRATED CMOS WIRELESS RECEIVERS
Leroux and steyaert
Vol. 843, ISBN: 1-4020-3190-4
SYSTEMATIC MODELING AND ANALYSIS OF TELECOM FRONTENDS AND THEIR BUILDING BLOCKS
Vanassche, Gielen, Sansen
Vol. 842, ISBN: 1-4020-3173-4
LOW-POWER DEEP SUB-MICRON CMOS LOGIC SUB-THRESHOLD CURRENT REDUCTION
Van der Meer, van Staveren, van Roermund
Vol. 841, ISBN: 1-4020-2848-2
WIDEBAND LOW NOISE AMPLIFIERS EXPLOITING THERMAL NOISE CANCELLATION
Bruccoleri, Klumperink, Nauta
Vol. 840, ISBN: 1-4020-3187-4
CMOS PLL SYSTHESIZERS: ANALYSIS AND DESIGN
Shu, Keliu, Sánchez-Sinencio, Edgar
Vol. 783, ISBN: 0-387-23668-6
SYSTEMATIC DESIGN OF SIGMA-DELTA ANALOG-TO-DIGITAL CONVERTERS
Bajdechi and Huijsing
Vol. 768, ISBN: 1-4020-7945-1
OPERATIONAL AMPLIFIER SPEED AND ACCURACY IMPROVEMENT
Ivanov and Filanovsky
Vol. 763, ISBN: 1-4020-7772-6
STATIC AND DYNAMIC PERFORMANCE LIMITATIONS FOR HIGH SPEED D/A CONVERTERS
Van den Bosch, Steyaert and Sansen
Vol. 761, ISBN: 1-4020-7761-0
DESIGN AND ANALYSIS OF HIGH EFFICIENCY LINE DRIVERS FOR Xdsl
Piessens and Steyaert
Vol. 759, ISBN: 1-4020-7727-0
LOW POWER ANALOG CMOS FOR CARDIAC PACEMAKERS
Silveira and Flandre
Vol. 758, ISBN: 1-4020-7719-X
MIXED-SIGNAL LAYOUT GENERATION CONCEPTS
Lin, van Roermund, Leenaerts
Vol. 751, ISBN: 1-4020-7598-7

CALIBRATION TECHNIQUES IN NYQUIST A/D CONVERTERS

By

Hendrik van der Ploeg

Philips Research,
Eindhoven, The Netherlands

and

Bram Nauta

University of Twente,
Enschede, The Netherlands

Springer

A C.I.P. Catalogue record for this book is available from the Library of Congress.

ISBN-10 1-4020-4634-0 (HB)
ISBN-13 978-1-4020-4634-6 (HB)
ISBN-10 1-4020-4635-9 (e-book)
ISBN-13 978-1-4020-4635-9 (e-book)

Published by Springer,
P.O. Box 17, 3300 AA Dordrecht, The Netherlands.

www.springer.com

Printed on acid-free paper

Printed in the Netherlands

Aan Ria en mijn ouders

Table of contents

List of abbreviations

A/D	analog-to-digital
ADC	analog-to-digital converter
AMP	amplifier
ATV	analog television
CCD	charge-coupled device
CDS	correlated double sampling
CMOS	complementary metal oxide semiconductor
CVBS	composite video baseband signal
D/A	digital-to-analog
DAC	digital-to-analog converter
DC	direct current
DNL	differential non-linearity
DSP	digital signal processor
DTV	digital television
EDGE	enhanced data rates for GSM evolution
ENOB	effective-number-of-bits
EPROM	electrically-programmable read-only memory
ERBW	effective resolution bandwidth
GSM	global system for mobile communication
IC	integrated circuit
INL	integral non-linearity
INT	integrator
LSB	least-significant-bit
MSB	most-significant-bit
MSCC	mixed-signal chopping and calibration
OPAMP	operational amplifier
OR	over-range
PCB	printed circuit board
PROM	programmable read-only memory

RAM	random-access memory
RF	radio frequency
ROM	read-only memory
SAR	successive approximation
SFDR	spurious free dynamic range
S/H	sample-and-hold
SHA	sample-and-hold amplifier
SINAD	signal-to-noise and distortion
SiP	system-in-package
SNR	signal-to-noise ratio
SoC	system-on-a-chip
SUB	subrange
T/H	track-and-hold
THA	track-and-hold amplifier
THD	total harmonic distortion
UR	under-range
VGA	variable gain amplifier
XOR	exclusive or
YUV	luminance and chrominance signals

List of symbols

a	ratio between f_{in}, f_s and f_c	-
A	gain factor	-
A_C	capacitor mismatch process parameter	$\sqrt{C}$
A_{fa}	folding amplifier gain	-
A_{V_T}	threshold mismatch process parameter	mVμm
$A_{\beta\square}$	current factor mismatch process parameter	%μm
C	symbol for capacitance	F
C_{gs}	gate-source capacitance	F
C_{intr}	minimum required capacitance	F
C_{LOAD}	load capacitance	F
$C_{matching}$	capacitance resulting from matching requirement	F
$C_{minimum}$	minimum required capacitance	F
C_{noise}	capacitance resulting from noise requirement	F
$C_{ox}^{\square}$	oxide capacitance per unit area	F/m^2
C_p	parasitic capacitance	F
C_x	capacitance determined by noise or matching requirements	F
DR_{pp}	peak-to-peak signal to rms noise	dB
$ENOB, DC$	effective-number-of-bits at DC	-
f_{-3dB}	–3 dB frequency	Hz
F_f	folding factor	-
F_{int}	interpolation factor	-
FoM	figure-of-merit	pJ
$FoM_{V_{pp}}$	figure-of-merit normalized to 1 V_{pp}	V^{2}pJ
f_c	chop signal frequency	Hz
$f_{ENOB,DC-0.5}$	frequency with 0.5 ENOB loss in ENOB w.r.t. $ENOB, DC$	Hz
f_{in}	input signal frequency	Hz
f_s	sample frequency	Hz

g_m	transconductance	A/V
G_R	gain ratio	-
I_{ds}	drain-source current	A
k	Boltzmann's constant, $1.3805 \cdot 10^{-23}$	J/K
K	circuit implementation factor	V^2
L	channel length of a MOS transistor	μm
$M(n)$	integrator content at sample n	-
$mult$	circuit multiplication factor	-
N	number of bits	-
N_C	number of coarse bits	-
NEF	noise excess factor	-
N_F	number of fine bits	-
N_{int}	number of interpolation stages	-
N_{linear}	number of pre-amplifier stages in linear region	-
N_{peak}	number of chopping frequencies	-
$N_{preamps}$	number of pre-amplifiers	-
offset red	offset reduction	-
#ORC	number of over-range comparators	-
p	number of out-of-range pre-amplifiers	-
P	power	W
$P_{noise,tot}$	total integrated noise power	W
$P_{noise,tot,circuit}$	total integrated noise power of a circuit	W
$P_{noise,tot,MOS}$	total integrated noise power of a MOS transistor	W
P_x	power determined by noise or matching requirements	W
R	symbol for resistance	Ω
s	(effective) oxide scaling factor	-
S_{v^2}	spectral noise power density	V^2/Hz
T	temperature	K
t_{ox}	(effective) oxide thickness	m
t_s	settle time	s
$t_{s,min}$	minimum sample period	s
$V_{amp,choppeak}$	amplitude at chop frequency	V
V_{dd}	supply voltage	V
$V_{full\ scale}$	full scale input signal	V
V_{gs}	gate-source voltage	V
V_{gt}	gate overdrive voltage	V
V_{LSB}	LSB voltage	V
$V_{noise,rms}$	rms value of the noise voltage	V

V_{offset}	offset voltage	V
V_{pp}	peak-to-peak voltage	V
V_{rms}	rms value of the voltage	V
$V_{shift,n}$	n^{th} reference voltage	V
$V_{signal,rms}$	rms value of the signal voltage	V
V_T	threshold voltage	V
W	channel width of a MOS transistor	μm
x	ratio between coarse comparator and residue amplifier offset	-
x_{NF}	normalizing factor for equal data rate	-
$\beta^{\square}$	current factor	μA/V^2
ΔC	error in the capacitance	F
ε	settling error	-
$\varepsilon_{coarse\ quantization}$	coarse quantization error	-
$\varepsilon_{fine\ quantization}$	fine quantization error	-
$\varepsilon_{fine\ range}$	fine range error	-
ε_{gain}	gain error	-
ε_{offset}	offset error	-
$\varepsilon_{reference}$	reference error	-
$\varepsilon_{subtraction}$	subtraction error	-
γ	white noise factor	-
$\mu_{n,p}$	mobility of electrons, holes	cm^2/Vs
π	pi, 3.141593	-
σ	standard deviation	-
σ_C	spread of the capacitance	F
$\sigma_{V_{offset}}$	spread of the offset voltage	V
τ	time constant	s
τ_{min}	minimum achievable time constant	s
τ_{unit}	time constant of one buffer unit	s

Preface

The advances in Integrated Circuits brought us advanced electronic systems available for large groups of people. By putting more and more functionality on an integrated circuit (IC) these systems could become cheap in mass production. Many researchers work on the advances in integration of complex systems.

Today, most electronic systems process signals in digital format. This way low-cost accurate circuits with a very high density can be made. The signals in the real world however are all analog, and therefore analog-to-digital and digital-to-analog converters are required. It is beneficial to integrate these A/D and D/A converters in the same IC as the digital functions for cost and size reasons. Since the digital ICs are generally implemented in CMOS technology, this requires A/D and D/A converters in CMOS as well, which poses quite some challenges in the design since CMOS technology is optimized for digital circuits.

Analog-to-digital converters can roughly be split in two families. The first family is the one of the high accuracy and low speed converters, often referred to as oversampled converters, typically used for audio and low MHz range. In these converters the speed of the technology is exploited to relax the accuracy demands of the analog circuits. The second family is the one of the high speed, medium accuracy converters, also referred to as Nyquist converters. Here oversampling cannot be used and the accuracy has to come from the analog circuits itself. Accuracy is generally achieved by relying on matching of equally designed on-chip components, which works better for larger-size components on a chip. However if large components have to work at high speed, this requires high power dissipation. So these converters generally have a trade-off in power, speed and accuracy.

This book deals with Nyquist-type converters. The basic idea exploited here is to use the available digital processing power of the CMOS technology to calibrate critical analog parts of the A/D converter. This way the analog accuracy demands of the components can be relaxed and smaller sized-components can be used, resulting in a reduction in power consumption. This book starts with an introduction in the field, exploring the power, accuracy and speed space. The conclusion is that two-step converters form a good base for investigating the calibration techniques.

In the remaining chapters several calibration algorithms are described, and several IC realizations of calibrated two-step converters are presented.

The book was originally the PhD thesis of Hendrik van der Ploeg who wrote it after 9 years of experience in A/D converter design at Philips Research laboratories. I really enjoyed working with Hendrik to prepare his thesis and now I feel very happy that it has been published as a book. I believe this book is really worth reading for a broad group of scientists and engineers.

Bram Nauta,
Professor,
University of Twente,
The Netherlands

Enschede, January 2006

Chapter 1

Introduction

1.1 A/D conversion systems

In modern systems, most of the signal processing is performed in the digital domain. Digital circuits have a lower sensitivity to noise and are less susceptible to fluctuations in supply and process variations. Unlike with analog circuits, signal processing in the digital domain offers greater programmability, error correction and storage possibilities. Since the world around us is analog and humans perceive information in the analog form, analog-to-digital (A/D) and digital-to-analog (D/A) converters represent important building blocks. They are found in many systems that require digital signal processing. This book focuses on A/D converters. A/D converters can be classified into two groups. There are A/D converters with a high accuracy and a low sample rate and A/D converters with a low accuracy and a high sample rate. This is illustrated in figure 1.1.

The first group includes sigma-delta converters for audio, signal transmission and instrumentation systems, while the second group includes video, camera and wide-band signal transmission systems. In order to increase the accuracy or the speed specifications of the A/D converters in both groups more power is required. The A/D converters from the second group of converters are dealt with in this book. They are found in products like television sets, security cameras, medical imaging devices, instrumentation, etc. The sampling speed required for these applications is generally in excess of 25 MSample/s and the resolution is 10 bits or more. A few examples of these applications are shown in figure 1.2.

The position of the A/D converter in such systems is shown in figure 1.3. In this figure, the signal is conditioned in the analog domain before it is applied to the A/D converter.

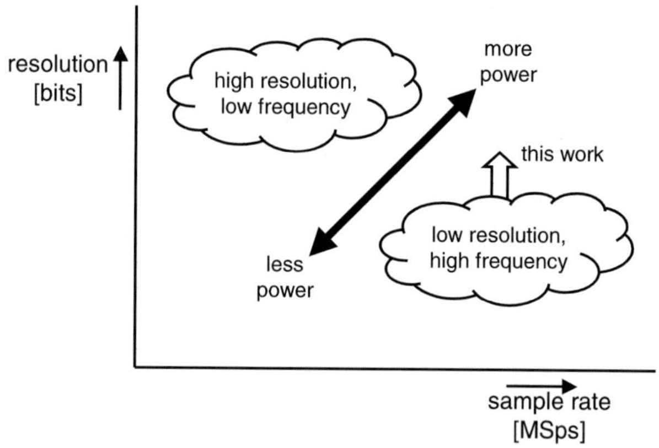

Figure 1.1: *High accuracy, low speed A/D converters and low accuracy, high speed A/D converters and the position of this work*

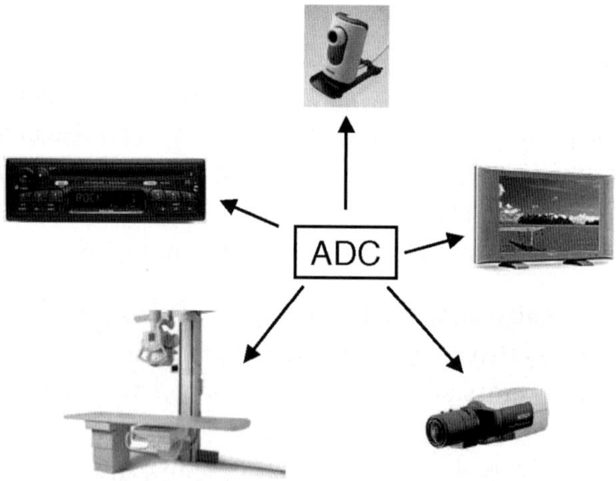

Figure 1.2: *Systems with A/D converters with sampling speeds in excess of 25 MSample/s and a resolution of 10 bits or more*

As shown in figure 1.3, the A/D converters form the connection between the analog world and the digital signal processing and digital memory. For example, the A/D converter converts the down-converted radio frequency (RF) antenna signal to the digital domain. In the case of figure 1.3, the filtering and channel selection is performed in the analog domain. Another example is an analog video signal with an aspect ratio of 4:3, which is converted to the digital domain. In the digital domain a field memory and additional processing is used to resize the video signal to an aspect ratio of 16:9. A D/A converter converts this signal back to the analog domain to be applied to a display [1]. Similar signal processing is required

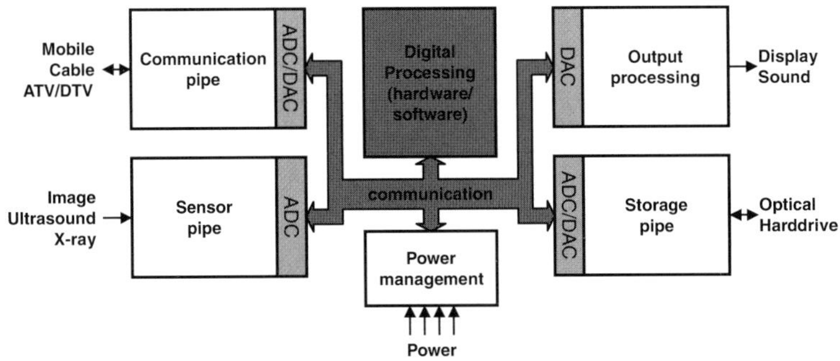

Figure 1.3: *The position of the A/D converter in the system, with signal conditioning in the analog domain*

to convert a video signal with a 50 Hz frame rate to a video signal with a 100 Hz frame rate.

The performance level should be such that the system is not affected by the imperfections of the data converter. Its design is therefore extremely important. Because of the trend towards decreasing feature sizes on silicon, it is becoming cheaper to shift analog functions, such as amplifying, filtering and mixing, into the digital domain. This involves shifting the A/D converter towards the input of the system; an extreme example of this is shown in figure 1.4.

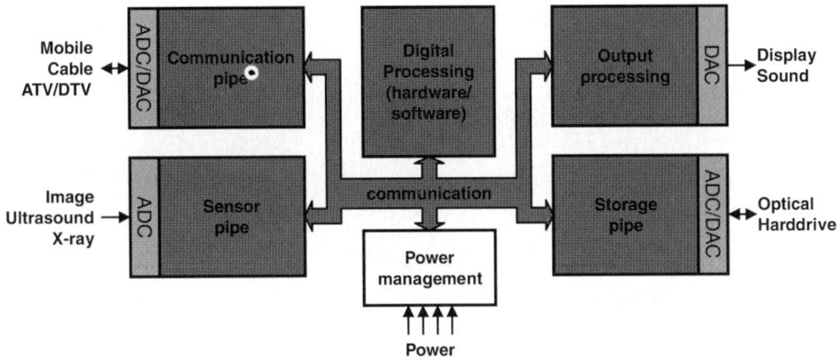

Figure 1.4: *The A/D converter shifted towards the input of the system*

In order to shift the A/D converter towards the input of such systems, A/D converters are required with a greater dynamic range and higher sampling speeds because there is less analog signal conditioning. This makes even higher demands on the A/D converter and potentially increases the power consumption. The calibration techniques investigated in this book enables to increase the accuracy without increasing the power consumption of the A/D converter.

If the specifications are known from the application, the challenges in the design of A/D converters arise from the technology used. The application generally determines the integrated circuit (IC) technology, which is mainly a cost-driven choice. Due to the high level of integration of systems on a chip, the digital functionality and therefore the area occupied by digital circuitry becomes dominant. This means a technology has to be chosen that is optimized for digital circuitry. In this case the technology is optimized for high-density digital circuits, which allows the use of small feature sizes that achieve a high packing density. The parameters of this technology are typically optimized for the digital circuitry and are therefore less suitable for high-performance analog circuits. On the other hand, there are technologies that are better suited for the design of dedicated high-performance analog circuits. Although the stand-alone analog circuits can achieve a high performance, these components have to be integrated into the overall system at a higher, system-in- package (SiP) level. This is in contradiction with the increasingly higher level of integration of systems-on-a-chip (SoC). However, the use of multi-die packages as shown in figure 1.5 also offers advantages because the digital circuitry can be designed on a digital chip, in a dedicated digital complementary metal oxide semiconductor (CMOS) technology, whilst the analog circuits are designed in a dedicated analog technology. This allows fast scaling of the digital part whilst the analog performance with the analog chip is maintained and digital cross-talk is prevented.

Figure 1.5: *System-in-package (from [2]) with analog TV processor and a digital signal processor*

A/D converter calibration techniques are applicable for any technology, however this book focuses on the design of A/D converters for highly integrated systems. The focus is, therefore, on designing A/D converters in CMOS.

1.2 Motivation and objectives

The main focus of this book is on improving the supply power efficiency of A/D converters capable of handling input signal frequencies up to the Nyquist frequency by using calibration. The boundary condition is to use standard CMOS technology without additional options. This allows us to use the A/D converter in a system-on-a-chip with a high level of integration. This book is split up into three main subjects:

- General analysis of the relation between the specified accuracy and speed and the resulting power for A/D converters.
- Comparison of the various A/D converter architectures with respect to achievable power efficiency.
- Investigation of enhancement techniques instrumental in increasing the level of accuracy whilst maintaining the power consumption of A/D converters.

The results from these investigations are used in the circuit design and realization of three A/D converters.

1.3 Layout of the book

In chapter 2 the relation between the required accuracy and speed and the power consumption in the design of A/D converters is explained. The minimum required capacitance, whether determined by matching or noise requirements, will be derived. This capacitance is charged and discharged at a certain speed. The calculated capacitance combined with the speed is a measure of the power consumption. This relation is also investigated over different CMOS technology generations.

For a given accuracy, the total minimum required capacitance of an A/D converter is a strong function of the architecture used. Chapter 3 describes the A/D converter architectures which are applicable for the scope of this book. The total minimum required capacitance is derived for each architecture and this result is used to compare the power consumption of these architectures. A benchmark has been carried out on recently published A/D converters and this is compared to the theoretical derivation. It is on the basis of this comparison that the two-step A/D converter architecture has been selected for further investigation.

In subranging A/D converters, matching of the ranges of the different quantization stages is very important. Chapter 4 presents dual-residue signal processing [3] in order to relax this requirement. The next part of chapter 4 deals with calibration techniques for reducing the matching requirements and the resulting minimum required capacitances. Different calibration techniques are discussed and the theory of a developed mixed-signal chopping and calibration (MSCC) algorithm is presented.

The design and the measurement results of a 10-bit two-step A/D converter are presented in chapter 5. In this design the dual-residue signal processing together with an analog offset compensation technique is demonstrated. The mixed-signal chopping and calibration algorithm is demonstrated in chapter 7 by means of a 12-bit two-step A/D converter. Chapter 7 shows the extension of the two-step to a 16-bit three-step converter in order to reduce the power consumption of the quantization stages. This design uses two independent mixed-signal chopping and calibration blocks.

In chapter 8 the main conclusions are summarized and recommendations for further research are presented.

Chapter 2

Accuracy, speed and power relation

2.1 Introduction

An A/D converter consists of several building blocks. Each of these building blocks has accuracy and speed requirements which are A/D converter architecture dependent. This chapter derives the relation between the accuracy, the speed and the resulting power consumption. This makes it possible to make a comparison between various A/D converter architectures based on the required accuracy and speed of the different building blocks.

The static (DC) accuracy of such building blocks is determined by the matching of components, which is technology dependent. The noise generated in a circuit sets a limit on the achievable 'dynamic' accuracy. In analog circuit design there are two main sources of noise: noise that arises as a result of a non-ideal environment (such as supply or ground noise) and noise from passive and active electronic devices. These sources of noise reduce the quality of the analog output signal of the circuit. Here only the electronic device noise is considered. In general, both mismatch and noise determine the quality of the conversion of a signal from the analog domain to the digital domain, assuming accurate timing. Section 2.2 shows that in order to guarantee a certain degree of accuracy in a given technology, the required conversion quality results in minimum device sizes and therefore minimum required capacitances. This minimum required capacitance together with the conversion speed is a measure for the minimum required power consumption, which is described in section 2.3. The minimum required capacitance is connected to a buffer which has its own parasitic capacitance. When this parasitic capacitance is larger than the load capacitance, the speed is limited by this parasitic capacitance.

The maximum achievable speed is discussed in section 2.4. Since the parameters that determine the accuracy and the speed are technology dependent, some trends in CMOS technologies are discussed in section 2.5.

2.2 IC-technology accuracy limitations

An A/D converter consists of non-ideal components. The specifications of the A/D converter can be translated into the accuracy demands of the respective components. The non-idealities are caused by both component spread determined by IC-processing and physical limits. In section 2.2.1 the effect of the IC-processing on component spread is described. Noise sets the limit for the smallest signal that can be processed with a specified quality. The effect of device noise on A/D parameters is discussed in section 2.2.2. Matching and noise demands result in a minimum required capacitance, which sets the minimum required power. This result is used in the next chapter to compare different A/D converter architectures.

2.2.1 Process mismatch

Due to spread in the IC-processing, nominally identical devices have limited matching. This mismatch is caused by differences in doping concentration and lithographic deviations on an atomic scale between two devices with identical layout. The effect of mismatch on the drain current in two identical devices can be calculated by using the expression that approximates the drain current of a CMOS transistor in saturation:

$$I_{ds} = \frac{\beta^{\square}}{2}\frac{W}{L}\left(V_{gs} - V_T\right)^2, \tag{2.1}$$

with $\beta^{\square} = \mu_{n,p}C_{ox}^{\square}$. Many spread mechanisms in the manufacture of transistors (such as: distribution of implants, local mobility fluctuations, oxide granularity, oxide charges, etc.) cause spread of transistor parameters. On a small area these processes have approximately a normal distribution. If a transistor consists of a large number of such - uncorrelated - small areas, the variation caused by these processes is a normal distribution with a spread which is proportional to $\frac{1}{\sqrt{WL}}$. These variations cause the current factor $\beta^{\square}$ and the threshold voltage V_T of two identical transistors to differ from each other [4, 5]:

$$\frac{\sigma\left(\Delta\beta^{\square}\right)}{\beta^{\square}} = \frac{A_{\beta^{\square}}}{\sqrt{WL}} \tag{2.2}$$

$$\sigma\left(\Delta V_T\right) = \frac{A_{V_T}}{\sqrt{WL}} \tag{2.3}$$

A_{V_T} is technology dependent and is proportional to the (effective) gate oxide thickness (t_{ox}) [5]. $A_{\beta^\square}$ is supposed to be determined by local mobility variations. These have a low dependence on processing and are therefore almost constant for different technologies [4].

Using equations 2.2 and 2.3 the V_T mismatch ($\sigma\left(V_{offset}\right)$) in two identical devices with the same operation conditions can be calculated:

$$\sigma\left(V_{offset}\right) = \sqrt{(\sigma\left(\Delta V_T\right))^2 + \left(\frac{V_{gs} - V_T}{2} \cdot \frac{\sigma\left(\Delta\beta^\square\right)}{\beta^\square}\right)^2} \tag{2.4}$$

For values for $V_{gt} = V_{gs} - V_T$, which are generally used below 200 mV to 300 mV, the V_T mismatch is dominant over the $\beta^\square$ mismatch [4]. This reduces equation 2.4 to:

$$\sigma\left(V_{offset}\right) = \frac{A_{V_T}}{\sqrt{WL}} \tag{2.5}$$

To show the effect of mismatch between two transistors on the accuracy of an A/D converter, the input differential pair of, for example, a comparator or pre-amplifier (figure 2.1) is considered. The offset of this input pair can be calculated using equation 2.5.

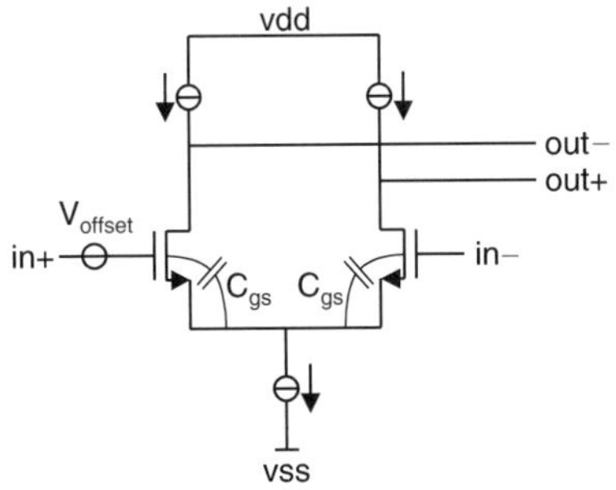

Figure 2.1: *Simple differential input pair with mismatch*

The differential input capacitance is mainly determined by the series connection of two gate source capacitors C_{gs} that equals $\frac{C_{gs}}{2}$. The required matching is a function of the area of the transistors used. Since the differential input capacitance also depends on this area it is derived that the input capacitance of a differential pair is dependent on the required matching:

$$C_{matching} = \frac{C_{gs}^\square}{2}\left(\frac{A_{V_T}}{\sigma\left(V_{offset}\right)}\right)^2 \tag{2.6}$$

For a transistor in saturation the capacitance per unit area is approximated by $C_{gs}^{\square} = \frac{2}{3}C_{ox}^{\square}$. The gate-drain capacitance is excluded for simplicity. This means that when the required matching of an input pair of a comparator is increased, the input capacitance increases by the square of this increase in accuracy. Equation 2.6 enables the calculation of the minimum intrinsic capacitance for a given accuracy requirement of V_{offset}.

2.2.2 Thermal noise

The random motion of electrons in a conductor causes a noise voltage across the conductor. This voltage is proportional to the absolute temperature. The generated squared voltage of a resistor with resistance R has a spectral density of:

$$S_{v^2}(f) = 4kTR,\ f \geq 0, \tag{2.7}$$

where $k = 1.38 \times 10^{-23}$ J/K is the Boltzmann constant. The spectral density in equation 2.7 gives the squared noise voltage per Hz. The noise power density is flat for the whole frequency spectrum, but because each conductor is connected to some kind of capacitance (for example a parasitic capacitance or the hold capacitor of a sample-and-hold circuit) the bandwidth of the noise is limited to the RC time constant of the resulting circuit. The total integrated noise power at the output of an RC circuit is:

$$P_{noise,tot} = \frac{kT}{C}, \tag{2.8}$$

which is independent of the resistance R. If the resistor in equation 2.8 is replaced by a MOS transistor connected as a common drain configuration, the total integrated noise power becomes:

$$P_{noise,tot,MOS} = \gamma\frac{kT}{C} \tag{2.9}$$

The factor γ equals $\frac{2}{3}$ for a long channel ($> 2\mu$m) device in saturation and has a value between one and two for short channel devices [6] in saturation.

If the conductor is not a single component as in equation 2.8 and equation 2.9, but a circuit consisting of more elements (for example a unity gain buffer), the total integrated output noise power becomes:

$$P_{noise,tot,circuit} = NEF\frac{kT}{C}, \tag{2.10}$$

when γ is set at one. The factor NEF represents the noise excess factor [7]. This is the ratio between the noise of the respective circuit and a single transistor. The

NEF is determined by the actual circuit implementation. Equation 2.10 can be rewritten as an rms value of the noise voltage:

$$V_{noise,rms} = \sqrt{NEF\frac{kT}{C}} \tag{2.11}$$

When the input signal is supposed to be a sine wave with an rms value of $V_{signal,rms} = \frac{V_{pp}}{2\sqrt{2}}$ and the required signal-to-noise ratio (SNR) is known from the A/D converter specification, the absolute minimum required capacitance to achieve this can be calculated:

$$C_{noise} = \frac{8 \cdot NEF \cdot 10^{\frac{SNR[dB]}{10}} kT}{V_{pp}^2} \tag{2.12}$$

Equation 2.12 shows that when the required accuracy increases by a factor of 2, which means the required SNR is increased by 6 dB, the required capacitance increases by a factor of 4.

2.2.3 Matching versus noise requirements

For a given circuit the component sizes are determined by the matching demands and the noise demands and the minimum required capacitances can be calculated. When the capacitance due to matching requirements is dominant, no additional capacitor is required to fulfil the total integrated noise requirement. However, when matching is not critical, which is the case in a track-and-hold amplifier (THA), for example, the minimum required capacitance $C_{minimum}$ is determined primarily by the noise requirements:

$$C_{minimum} = MAX\left(C_{matching}, C_{noise}\right) \tag{2.13}$$

This minimum required capacitor $C_{minimum}$ can be used to calculate the required power. This is discussed in the next section.

2.3 Speed and power

For an A/D converter to operate with sufficient accuracy, the error made during quantization needs to be sufficiently small. For a Nyquist A/D converter the implemented number of bits determines the quantization error. To limit further degradation of the input signal, the errors added due to implementation, like mismatch errors, noise and settling errors, need to be sufficiently small. The effects of mismatch errors and noise are dependent on the actual A/D converter implementation and are discussed in chapter 3. Another important source of errors is settling (or dynamic) errors. The bandwidth or speed of the analog circuitry is generally dependent on accuracy and power. This section determines the relation between these parameters. First of all, sampled signals are considered. Then it is shown that if the analog signal does not exceed the Nyquist frequency, the sampled and non-sampled signals will have a similar speed, power and accuracy relation.

The conceptual circuit is shown in figure 2.2. This can, for example, be a unity gain buffer with input transconductance g_m. The settling errors are made when the load capacitor in figure 2.2 is not charged completely by the active circuit. This load capacitor can be determined by matching or noise requirements.

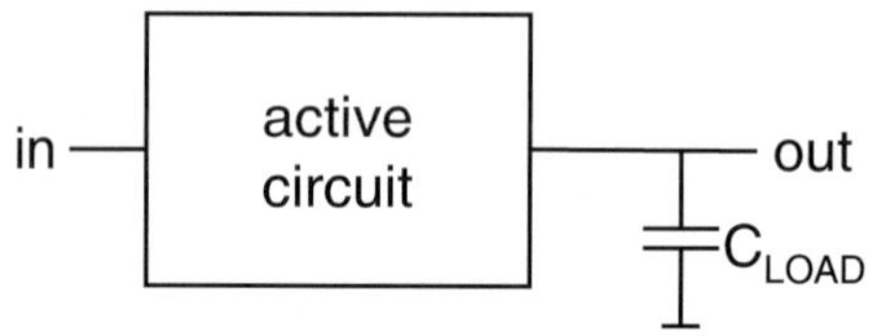

Figure 2.2: *An active circuit with an input transconductance g_m driving C_{LOAD}*

The circuit from figure 2.2 has to respond to a full range step signal with sufficient accuracy. The allowed settling error is determined by the A/D converter architecture, as will be discussed in chapter 3. The settling error is generally composed of a linear settling part and a slewing part. Only linear settling is considered here. However, the errors due to insufficient linear settling do not distort the signal but cause an attenuation of the signal (which is unwanted for some architectures). For a certain maximum available settling time t_s, which is a function of the sample rate of the A/D converter, a load capacitor C_{LOAD} and a maximum allowed settling error ε, the required transconductance g_m of the circuit in figure 2.2 is calculated by:

$$g_m = \frac{-C_{LOAD}\ln(\varepsilon)}{t_s} \tag{2.14}$$

The required power of the active circuit in figure 2.2 is proportional to the transconductance g_m of equation 2.14:

$$P = K \cdot g_m = K \cdot \frac{-C_{LOAD}\ln(\varepsilon)}{t_s}, \tag{2.15}$$

where K is determined by the actual circuit implementation and the technology. In equation 2.15, C_{LOAD} and ε are both dependent on the required accuracy, which is dependant on the A/D architecture. If these are increased, therefore, the power is increased by the same amount to maintain an equal settling time t_s. In this case the operating point condition of the active circuit in figure 2.2 remains constant and only the currents and widths of the transistors are scaled. The power is inversely proportional to the settling time t_s. However, when the parasitic capacitance of the active circuit becomes approximately C_{LOAD}, this dependency is no longer inversely proportional. This is discussed in the next section.

When the signal applied to the load capacitor is not sampled, as is the case in the input stage of a THA, for example, the circuit from figure 2.2 has to provide sufficient analog bandwidth. Suppose the available settling time t_s from equation 2.14 is equal to the sample period $\frac{1}{f_s}$ of a certain A/D converter. Then f_{-3dB} can be written, using equation 2.14, as:

$$f_{-3dB} = \frac{g_m}{2\pi C_{LOAD}} = -\frac{\ln(\varepsilon)}{2\pi t_s} = -\frac{\ln(\varepsilon)}{\pi}\frac{f_s}{2}, \tag{2.16}$$

From equation 2.16 it can be concluded that for $\varepsilon < 0.04$ (this equals a accuracy of $N < 4$ bit), the analog bandwidth (f_{-3dB}) is larger than the Nyquist frequency ($\frac{f_s}{2}$), when the settling time of the respective circuit equals $\frac{1}{f_s}$. For the sake of simplicity, in the comparison of the different A/D converter architectures in the next chapter the circuits working with the analog sampled and non-sampled signals are treated similarly.

The next chapter derives the total intrinsic capacitive load depending on the A/D converter architecture and the accuracy. To be able to make a comparison of the architectures described, it is assumed that the content of the active circuit in figure 2.2 and the technology are the same for all architectures, which means K in equation 2.15 is constant. This allows comparison of the minimum power required by the different architectures.

2.4 Maximum speed

The power needed to drive the capacitive load of the circuit from figure 2.2 is linearly dependent on the load when the parasitic capacitance of this circuit is negligible with respect to the load capacitance. In general this is the case if the required time constant τ is large. This is explained in the following equation:

$$\tau = \frac{C_{LOAD} + C_p}{g_m}, \tag{2.17}$$

with C_p as the parasitic capacitance. When the time constant τ is large, C_p is negligible with respect to C_{LOAD} and equation 2.17 can be rewritten to equation 2.14. However, when a smaller time constant is required, the g_m of the circuit needs to be increased. This is done by increasing the currents and the widths W of the transistors equally. For example, two equal circuits can virtually be put in parallel to achieve half the settling time while maintaining the same operating condition. This is shown in figure 2.3.

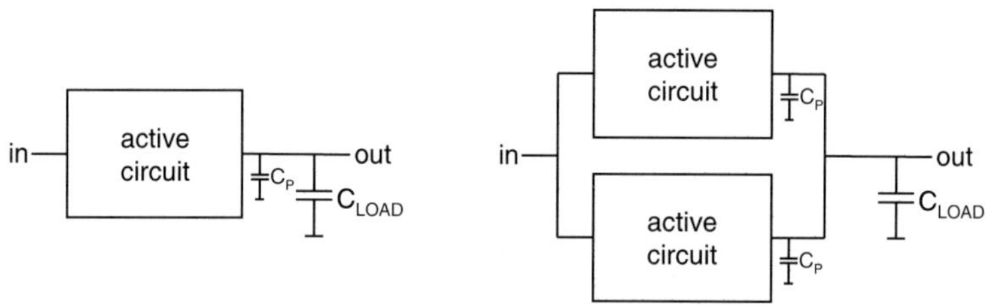

Figure 2.3: *Halving the time constant by doubling the circuit*

When two equal circuits are put in parallel, not only does the drive capability double, but the parasitic capacitance of the circuit doubles as well, as shown in figure 2.3. To reduce the time constant further, more circuits can be put in parallel. The buffer shown in figure 2.4 gives an example of this.

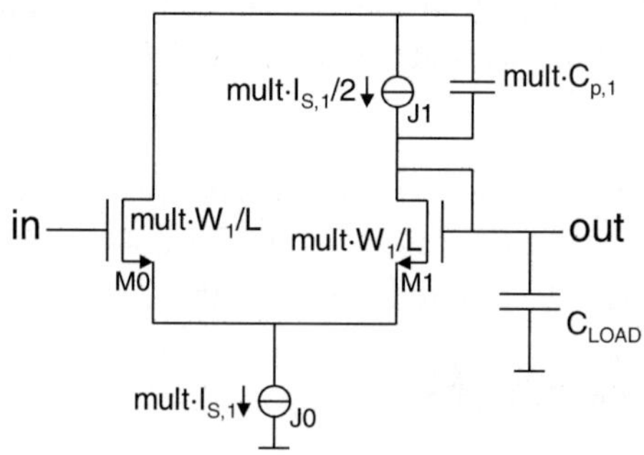

Figure 2.4: *Unit buffer to drive a C_{LOAD} with parasitic capacitance C_p*

The parasitic capacitance C_p models the junction capacitances of both $M1$ and the current source $J1$. Suppose the buffer drives a certain load capacitance C_{LOAD} and the time constant is equal to τ_{unit}. When the time constant needs to be decreased, the number of parallel buffers is increased by a factor $mult$. Figure 2.5 shows a simulation result of the time constant relative to τ_{unit} as a function of the number of parallel buffers $mult$.

If the parasitic capacitance is negligible with respect to the load capacitance (in this example at $mult = 1$), the time constant decreases linearly with the number of parallel circuits. For a larger number of parallel circuits (in this example beyond $mult = 10$), the parasitic capacitance is no longer negligible and the improvement in time constant is not linear. The minimum time constant is reached

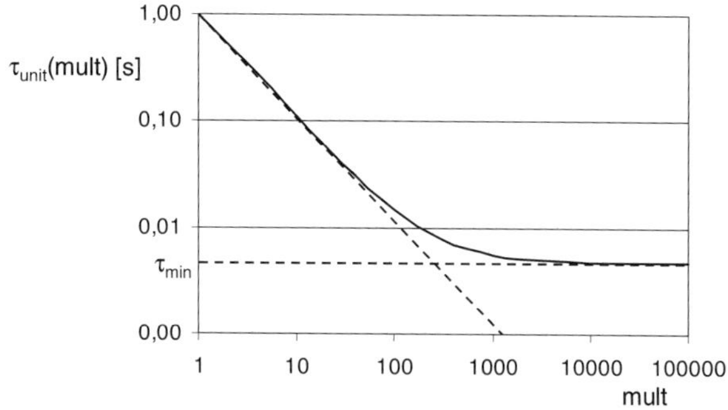

Figure 2.5: *Simulation of the relative time constant as a function of the number of parallel buffers*

when the parasitic capacitance is dominant over the load capacitor (see also equation 2.17). This means that no speed improvement is achieved when the power of the circuit is increased, since g_m and C_p increase equally. In general, this means that for a power-efficient implementation the parasitic capacitance should be smaller than the applied load capacitance, otherwise more than half of the output power of the circuit is consumed to drive its own parasitic capacitance. The minimum time constant τ_{min} which can be reached is independent of the applied load capacitance C_{LOAD} and is determined by the ratio of g_m and C_p of the actual circuit implementation.

2.5 CMOS technology trends

Important parameters which determine the matching and the allowed signal swing are technology dependent. The required load as derived earlier in this chapter is therefore determined very much by the technology used. The parameters which determine the required power and the minimum achievable settling time, such as the current factor $\beta^{\square}$, minimum gate length L and parasitic capacitances are also dependent on the technology.

Minimum power scaling

The oxide breakdown voltage and therefore the oxide material and thickness determines the permitted supply voltage and achievable signal swing. In this section the scaling factor for the (effective) oxide thickness (t_{ox}) and minimum gate length is assumed to be s. As a starting point, for a certain realization of a circuit in the 0.18 μm CMOS technology, s is set to one. s is smaller than one for more advanced technologies. With the use of equations 2.6 and 2.11, equations 2.18 and 2.19 derive the relative dependence of the minimum required capacitance for a matching and a noise-limited design as a function of s. It is assumed that A_{V_T} [5], $C_{ox}^{\square}$ and V_{rms} [8] are scaling with s and that NEF remains constant for different technologies.

$$C_{matching} = \frac{C_{ox}^{\square}}{3}\left(\frac{A_{V_T}}{\sigma(V_{offset})}\right)^2 \propto \frac{1}{s} \tag{2.18}$$

$$C_{noise} = \frac{NEF\,kT}{V_{noise,rms}^2} \propto \frac{1}{s^2} \tag{2.19}$$

Figure 2.6 shows the theoretical relative dependance of the minimum required capacitance for different technologies.

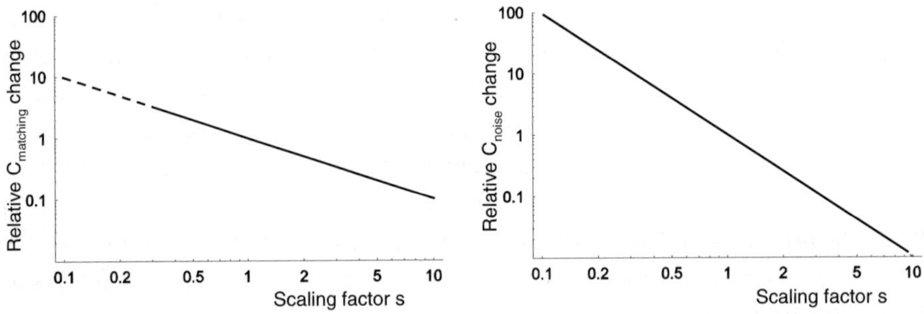

Figure 2.6: *The relative capacitance dependence on the scaling factor s*

The curve of $C_{matching}$ left side (for small s) is dashed, since it is dependent on technology parameters, which might not have the dependency on technology, as described above, for future technologies. As can be seen in figure 2.6, for matching- and noise-limited designs the capacitance increases with $\frac{1}{s}$ and $\frac{1}{s^2}$ respectively. To calculate the resulting power, it is assumed that the bandwidth (or the required settling time) remains constant, that V_{gt} scales linearly with V_{dd}, and that $\frac{g_m}{I_{ds}}$ is inversely proportional to V_{gt} (for the chosen operation point where V_{gt} is between 100 mV and 300 mV [8]):

$$\left.\begin{array}{l} P = V_{dd} \cdot I_{ds} \\ I_{ds} \propto V_{gt} \cdot g_m \overset{g_m \propto C_x}{\propto} V_{dd} \cdot C_x \end{array}\right\} \overset{V_{dd} \propto s}{\Rightarrow} P_x \propto C_x \cdot s^2 \tag{2.20}$$

In equation 2.20 C_x is $C_{matching}$ for a matching-limited design and C_{noise} for a noise-limited design. Figure 2.7 shows the resulting relative power dependence for a matching- and a noise-limited design as a function of the scaling factor s.

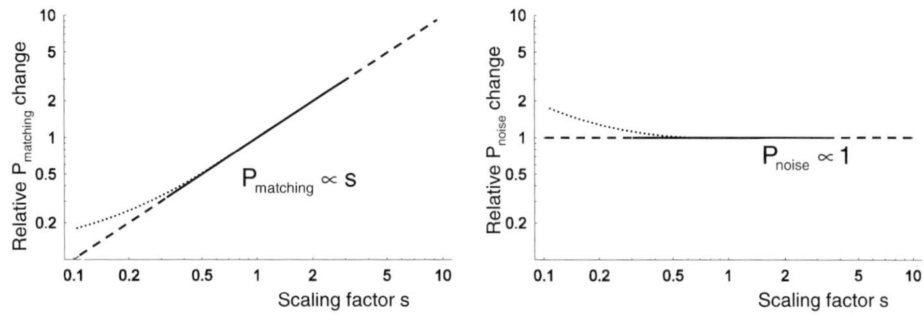

Figure 2.7: *The relative power dependence on the scaling factor s for constant bandwidth*

V_{gt} cannot be scaled down infinitely with s. For small V_{gt} (near sub-threshold), $\frac{g_m}{I_{ds}}$ is not inverse proportional to V_{gt}, but $\frac{g_m}{I_{ds}}$ becomes inversely proportional to $\sqrt{V_{gt}}$ [8]. In this case equation 2.20 changes into:

$$I_{ds} \propto \sqrt{V_{gt}} \cdot g_m \propto \sqrt{V_{dd}} \cdot C_x \Rightarrow P_x \propto C_x \cdot s^{\frac{3}{2}} \tag{2.21}$$

Due to the non-linear relation between $\frac{g_m}{I_{ds}}$ and $\frac{1}{V_{gt}}$ for small V_{gt} the power becomes approximately inversely proportional to the square root of s for a noise-limited design, while it is proportional to the square root of s for a matching-limited design. This is indicated in figure 2.7 with the dotted lines. This means that for low accuracy, high-speed A/D converters it is advantageous to move into a more advanced technology, while a high-accuracy A/D converter, which is generally noise limited, is more power efficient in an older technology (if the A/D converter architecture is kept constant).

An interesting situation occurs for advanced technologies ($s < 1$ in figure 2.7) when the supply voltage is kept constant and critical transistors are implemented in the most advanced, thin oxide transistors. The voltages on and between all transistor terminals then have to be sufficiently small. The non-critical transistors are implemented in thick-oxide technology. The critical transistors determine the g_m. In this case, the noise capacitance shown in figure 2.6 does not change as a function of s, while the matching dependent capacitance is proportional to s, since the input signal amplitude can be maintained. Because V_{dd} and therefore V_{gt} do not scale, the power for a noise-limited design will remain constant, but more importantly, the power for a matching-limited design will remain proportional to s. This, however, is not a standard design technique.

Maximum speed scaling

The parasitic capacitance determines the maximum achievable speed. Generally, the major contribution of the parasitic capacitance (C_p in figure 2.3) at the output of a circuit comes from the drain junction capacitor. This drain junction capacitance is relatively constant per unit width over several CMOS technologies [9]. However, due to scaling with s, the minimum L scales with s, $\beta^{\square}$ scales with $\frac{1}{s}$ and V_{gt} is scaled with s. Scaling of the quadratic equation for the drain current results in:

$$I_{ds} = \frac{\beta^{\square}}{2} \cdot \frac{W}{L} \cdot V_{gt}^2 \propto \frac{1}{s} \cdot \frac{W}{s} s^2 \propto W \propto C_p \tag{2.22}$$

It shows that I_{ds} is proportional to the parasitic capacitance, regardless of the scaling factor s. Using the relation between $\frac{g_m}{I_{ds}}$ and V_{gt}, the effect of scaling on the minimum time constant τ_{min} can be derived:

$$\frac{1}{\tau_{min}} \propto \frac{g_m}{C_p} \propto \frac{g_m}{I_{ds}} \propto \frac{1}{V_{gt}} \propto \frac{1}{s} \tag{2.23}$$

In equation 2.23 it is shown that the minimum time constant scales with s. This means that when using a more advanced technology, the maximum achievable speed is increased. In equation 2.23 it is assumed that V_{gt} is between 100 mV and 300 mV [8]. For smaller V_{gt}, $\frac{g_m}{I_{ds}}$ is inversely proportional to $\sqrt{V_{gt}}$. The relation between τ_{min} and s then becomes less than linear and the improvement in maximum speed when the design is transferred to a more advanced technology is less than for designs with $V_{gt} < 100$ mV.

If the most advanced transistors are used while the supply voltage is kept constant, V_{gt} can also be kept constant. This results in $\tau_{min} \propto \frac{1}{s^2}$.

2.6 Conclusions

The minimum required capacitances in an A/D converter are determined by the noise or the matching requirements. The minimum required capacitance from both the noise and the matching requirements are quadratically dependent on the required accuracy, while the total capacitance is determined by the maximum of the capacitance from the noise or matching requirement. This maximum capacitance needs to be charged with the respective signal with sufficient accuracy within a certain time period, which is a measure for the power consumption. The total required capacitance, the accuracy and the speed are A/D converter architecture dependent. This makes it possible to compare the power consumption of different A/D converter architectures, as discussed in chapter 3.

This comparison can be made because the power consumption is proportional to the load capacitance and inversely proportional to the available settling time, if the parasitic capacitor is smaller than the load capacitance. If the available settling time approaches its lower limit, which means that the parasitic capacitance reaches the load capacitance, the power consumption is no longer inversely proportional to the time constant, but increases dramatically.

In this chapter it is shown that in matching-limited designs it is beneficial to use the most advanced technology. The decreasing power supply reduces the allowed signal swing, however the matching parameter A_{V_T} also decreases. For noise-limited designs, the input signal amplitude should be as large as possible, which calls for older technologies with larger supply voltages. The maximum speed is achieved in the most advanced technologies where the ratio between g_m and the parasitic capacitance C_p generally increases.

The errors caused by matching errors can be reduced by using calibration techniques, which will be discussed in chapter 4. The capacitance resulting from matching requirements can therefore be decreased. However since noise is a random process, it can not be calibrated and the capacitance determined by noise requirements is the lower limit and can not be reduced by calibration.

Chapter 3

A/D converter architecture comparison

3.1 Introduction

A number of different A/D architectures exist for a given set of specifications. This chapter describes the main known architectures and their limitations in terms of minimum required capacitance, power and speed. The architectures can be broadly classified into parallel and serial structures. The former include flash, folding and two (or multi) step converter, where the decision about an input signal is made by a number of parallel devices (i.e. comparators). The second category includes pipe-line and successive approximation, where in general only one device is (re-)used to make the decision. Sigma Delta, counting and (dual) slope A/D converters are not taken into account in this comparison, since only wide-band or Nyquist converters are considered in this book.

The choice of technology is determined primarily by the application. There are two main possibilities in this respect. First is the class of stand-alone A/D converters. For this class, the specifications to be achieved by the A/D converter determine the technology. In this book the focus is on the second class, the embedded applications, where the analog part, which contains the A/D converter, needs to be integrated with the digital circuits on a single chip. For systems with large digital processing this means that the A/D converter has to be designed in a CMOS technology optimized for digital circuits. For example, the supply voltage in this case is reduced with newer technologies. A/D converters designed in CMOS technologies optimized for digital circuits have an additional requirement. They need to be robust with respect to scaling. Newer digital CMOS technologies allow reduction of the area of the digital circuitry, which is cost efficient. However, the analog circuits, like A/D converters, also have to operate in this newer technology (with reduced supply voltage). The matching parameters for transistors, resistors and capacitors that are determined by the technology are also very important in the design of A/D converters. The basic A/D converter architectures are described in this chapter, however publications about these architectures deal with more specific solutions to ensure the greatest possible independence from the limiting processing parameters.

The different A/D converter architectures can be compared with each other with respect to power efficiency because, together with the achieved accuracy and sampling speed, the efficiency of the architecture can be derived. In chapter 2 it is derived that the power is proportional to the capacitance that has to be charged. This capacitance is a function of the architecture. In the next sections the minimum capacitance required to perform a proper conversion is derived from the architecture. This is called the intrinsiccapacitance C_{intr}. This capacitance can be determined by noise or matching requirements. Together with the speed required to charge this capacitance, the figure-of-merit of the architectures is derived, enabling a proper comparison of the power efficiency for each architecture. This is done in section 3.7. In the comparison the power required in the digital part of the respective A/D converter is ignored for the sake of simplicity.

The architectures that are compared in this chapter are integrated circuit, Nyquist converters. These are:

- Flash
- Folding
- Two-Step / Subranging
- Pipe-line
- Successive approximation

Each section first describes the architecture and its accuracy-limiting components and from this the total intrinsic capacitance is calculated.

3.2 Flash

The most commonly known architecture is the flash A/D converter [10, 11, 12, 13, 14]. This architecture has several implementation possibilities. First the full-flash architecture is considered, and then two enhancement techniques are described: interpolation and averaging.

3.2.1 Full flash

Architecture

The most straightforward implementation of the flash A/D converter is the full-flash architecture (figure 3.1). This converter contains a comparator for each decision level. It is often used for very high-speed and low-accuracy applications [13, 14].

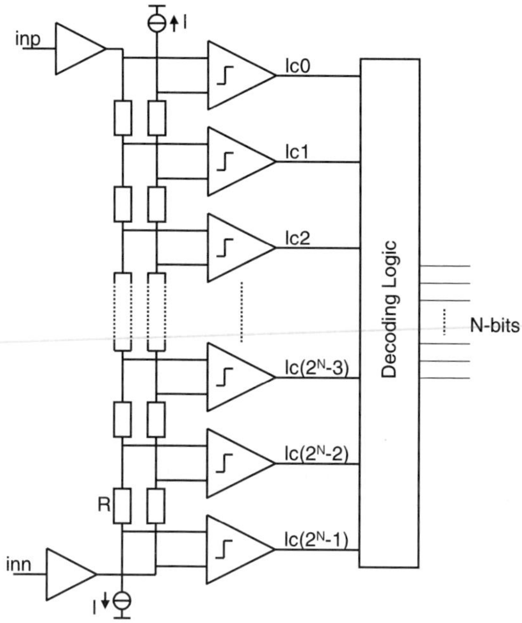

Figure 3.1: *A differential flash A/D converter*

In order to generate the zero-crossings of the A/D converter, the differential input signal is applied to the tops of two floating resistor ladders via two buffers. The voltages on the resistor taps on the ladders are the input voltage shifted with a certain reference ($V_{shift,n} = nR \cdot I$). Since each decision level of the A/D converter requires a comparator, $2^N - 1$ comparators are connected to the ladders to detect the zero-crossings. The output of the comparators is a thermometer code, the output is zero for all the comparators with an input signal smaller than zero, while the output is one for all the comparators with an input signal larger than zero. This thermometer code is decoded into a binary code by means of digital logic. Since the quantization is carried out at one point in time, the latency of such a converter is small - generally one to two clock cycles, which are necessary for comparison and decoding (not shown in figure 3.1). This makes the flash converter very suitable for high-speed feedback systems.

Accuracy

Since each level of the flash A/D converter has to be detected with the accuracy of the overall converter accuracy, each comparator making this decision should have this accuracy. The offset of the input stage of each comparator must therefore be sufficiently low (section 2.2.1). The number of bits N which have to be resolved by the A/D converter determines the accuracy of a decision level. In general, the accuracy of a decision level has to be within $\frac{1}{4}$ of a least-significant bit (LSB) (appendix A) for a degradation of less than 2.4 dB in signal-to-noise and distortion (SINAD).

Intrinsic capacitance

The intrinsic capacitance, which consists of the capacitive load of the comparators, is determined by two factors. The number of bits (N) of the A/D converter determines the number of comparators and the required comparator accuracy. The number of comparators is $2^N - 1$. Since the offset voltage of the comparators is assumed to be a Gaussian distribution with zero mean and a standard deviation σ, the probability that all comparators of the flash converter are within a certain limit can be calculated [1]. This probability corresponds to the yield of the A/D converter. The required $\sigma_{V_{offset}}$ of a single comparator is a function of this required yield. The condition that all the comparators must have an offset smaller than $\frac{1}{4}$ of an LSB results in the following equation for the yield of an N-bit flash converter:

$$Yield = \left(1 - P\left(\frac{V_{offset}}{\sigma} > \frac{LSB}{4\sigma}\right)\right)^{2^N-1} \tag{3.1}$$

The probability function P is in this case normalized to the standard normal distribution N(0,1). This equation can be used to derive the demands for $\sigma_{V_{offset}}$ if a yield of 0.99 is required, with a maximum offset of $\frac{1}{4}$ of an LSB. Figure 3.2 shows the required ratio $\frac{\sigma_{V_{offset}}}{V_{LSB}}$ to achieve a yield of 0.99, as a function of the number of bits (N).

As can be seen in figure 3.2, the difference between $\frac{\sigma_{V_{offset}}}{V_{LSB}}$ for $N = 1$ and $N = 20$ is only a factor two, while the difference between $N = 1$ and $N = 20$ in the number of comparators is already 2^{20}. Therefore, for the sake of simplicity for the calculation of the intrinsic capacitance the maximum allowed comparator offset voltage is assumed to have a constant value of 0.0625, for all N, which equals $\frac{1}{4}$ of $\frac{LSB}{4}$. The size of the input transistors is calculated using equation 2.6, with

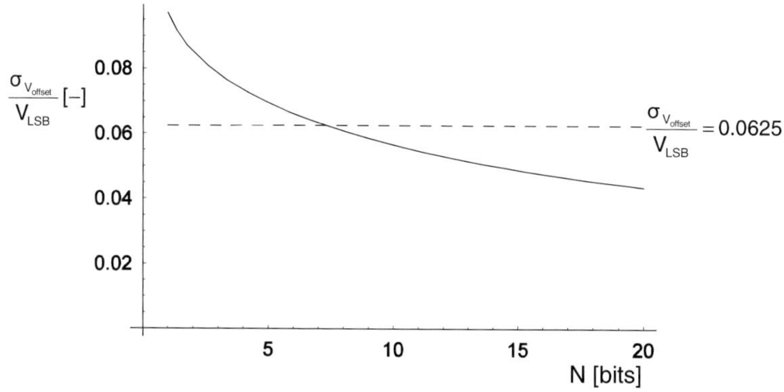

Figure 3.2: *Required* $\frac{\sigma V_{offset}}{V_{LSB}}$ *as a function of the number of bits (N) for a yield of 0.99, with a maximum offset of* $\frac{LSB}{4}$

$\sigma V_{offset} = \frac{V_{pp}}{4 \cdot 4 \cdot 2^N} = \frac{V_{pp}}{2^{N+4}}$. The total intrinsic capacitance of these comparators can then be calculated by:

$$C_{intr,flash} = (2^N - 1) \cdot C_{diff,input,matching} \approx \frac{C_{ox}^{\Box} A_{V_T}^2}{3V_{pp}^2} \cdot 2^{3N+8} \tag{3.2}$$

The total intrinsic capacitance is calculated in equation 3.2 assuming that the offset of the comparators is only determined by the input transistors. In practice, the circuitry after the input transistors also contributes to the offset, depending on the actual implementation. For the sake of simplicity, however, only the input transistors are considered in this analysis. If the number of bits in a flash converter is increased, the load increases by a factor of 8 for every extra bit. Figure 3.3 shows the total load capacitance as a function of N.

From figure 3.3 the total intrinsic capacitance of the flash A/D converter can be calculated. For example, for an 0.18μm technology: $A_{V_T} = 5$ mV, $C_{ox}^{\Box} = 7$ fF/μm^2 and $V_{pp} = 1$ V, for a 10-bit converter the intrinsic capacitance is over 16 nF! Because of the large amount of comparators needed at the overall accuracy, this capacitance is high for large N. Consequently, techniques to reduce the intrinsic capacitance of a flash A/D converter, like interpolation and averaging, have been developed. They are described in the next sections.

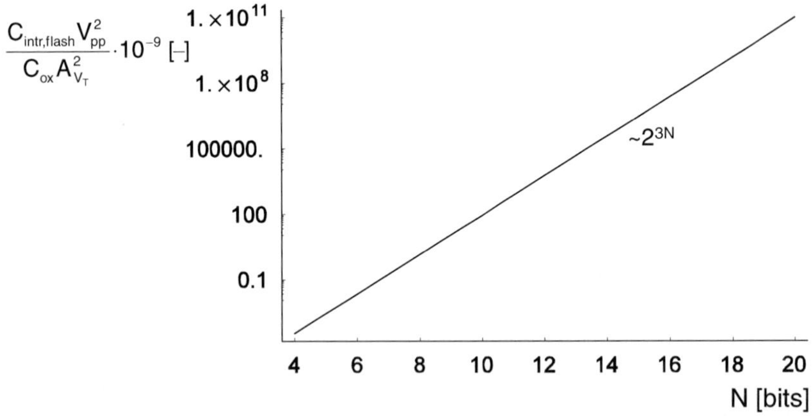

Figure 3.3: *Total intrinsic capacitance of a flash A/D converter as a function of the number of bits (N)*

3.2.2 Interpolation

Architecture

The required zero-crossings can be derived from the floating ladders directly as shown in figure 3.1. This requires $2^N - 1$ input pairs connected to the ladders. The $2^N - 1$ required zero-crossings can also be generated by interpolation [15] of pre-amplifier outputs as shown in figure 3.4a.

The pre-amplifiers with gain A in figure 3.4a generate the zero-crossings of the ladder taps. By taking the outputs of adjacent pre-amplifiers, additional zero-crossings are generated as shown in figure 3.4b. This interpolation substantially decreases the input capacitance seen at the ladder, since the number of input pairs is halved. The number of interpolation stages (N_{int}) can be more than one. When several interpolating pre-amplifier stages are used ($N_{int} > 1$) instead of only one pre-amplifier stage, the number of input pairs connected to the ladders is reduced even further. $2^N - 1$ comparators are still required for the quantization of the $2^N - 1$ zero-crossings. When gain is applied in the interpolating pre-amplifier stages, however, the accuracy requirements of these comparators are reduced. The intrinsic capacitance of the comparators is thus greatly reduced.

In practice, the transfer curve of a pre-amplifier is non-ideal. The active region of the input of a pre-amplifier, consisting of a differential pair, is limited by the limited linear region of this input transistor pair. When the linear region is not sufficient, the interpolated zero-crossings deviate from the wanted position [16].

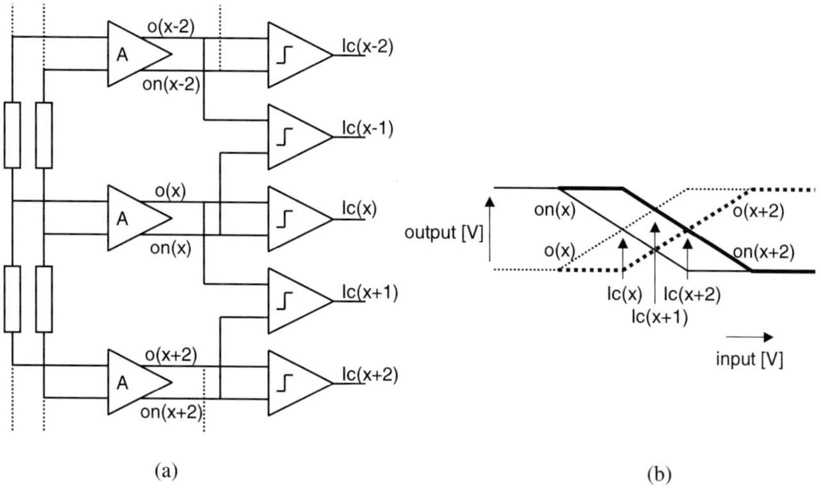

Figure 3.4: *Interpolation of pre-amplifier outputs (a) and additional zero-crossing generation (b) for one interpolation stage ($N_{int} = 1$)*

In this derivation it is assumed that the interpolation is ideal, which means that no additional power is required for a sufficiently large linear region.

Intrinsic capacitance reduction

The gain factor A of each stage reduces the accuracy requirements of the subsequent stages. For the sake of simplicity it is supposed that the gain factor A is equal to A_0 for all stages and the offset contribution referred to the input scales with A_0 per interpolation stage. The intrinsic capacitance of one input pair in each interpolation stage i can then be calculated as a function of the number of interpolation stages N_{int}:

$$C_{intr,pair,i} = \left(\frac{N_{int}}{A_0} + 1\right) \cdot \frac{C_{ox}^{\square} A_{V_T}^2}{3A_0^{2i} V_{pp}^2} \cdot 2^{2N+8}, \qquad (3.3)$$

for $0 \leq i \leq N_{int}$. In figure 3.4a it can be seen that the input capacitance of one pre-amplifier is given by equation 3.3 for $i = 0$, while the input capacitance of one comparator is given by equation 3.3 for $i = 1 = N_{int}$. With equation 3.3 the total intrinsic capacitance of an interpolated flash A/D converter can be calculated as a function of N_{int}, by summing the intrinsic capacitances $C_{intr,pair,i}$ of all the input pairs in each interpolation stage:

$$C_{intr,flash\ interpolation} = \sum_{i=0}^{N_{int}} (2^{N-(N_{int}-i)} - 1) C_{intr,pair,i}$$
$$\approx \sum_{i=0}^{N_{int}} 2^{N-(N_{int}-i)} C_{intr,pair,i} \quad (3.4)$$

To calculate the effect of interpolation on the intrinsic capacitance of a flash A/D converter equation 3.4 is divided by equation 3.2:

$$\frac{C_{intr,flash\ interpolation}}{C_{intr,flash}} \approx \sum_{i=0}^{N_{int}} \left(\frac{N_{int}}{A_0} + 1\right) \cdot \frac{2^{i-N_{int}}}{A_0^{2i}} \quad (3.5)$$

This ratio is only dependent on the number of subsequent interpolation stages N_{int} and the pre-amplifier gain A_0. Equation 3.5 is plotted in figure 3.5 as a function of A_0, with N_{int} as a parameter.

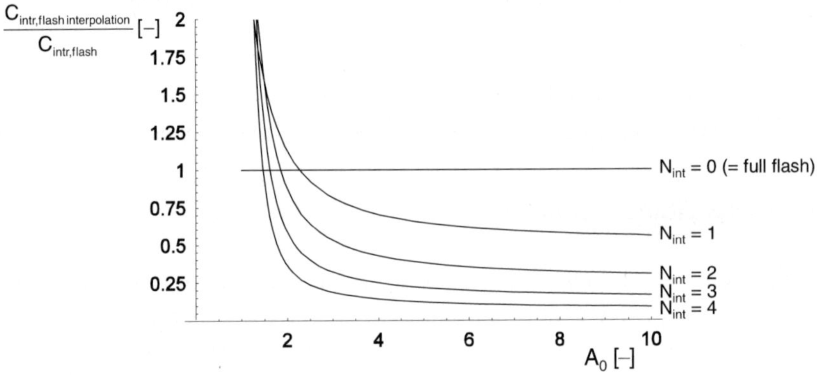

Figure 3.5: *Improvement in intrinsic capacitance by using interpolation as a function of the pre-amplifier gain A_0 with the interpolation factor N_{int} as a parameter*

When $N_{int} = 0$ the ratio of equation 3.5 equals one, since this is the same as the flash A/D converter without interpolation. This is independent of the gain A_0, since there is no pre-amplifier stage. For $N_{int} \geq 1$ and low gain A_0, the A/D converter with interpolation has more intrinsic capacitance than the flash A/D converter without interpolation. The reason for this is that with interpolation more stages contribute to the offset with respect to the flash A/D converter, which causes a higher intrinsic capacitance. For larger gain A_0, the effect of the gain is that the stages after the first interpolation stage contribute less to the total input-referred offset, resulting in a lower intrinsic capacitance. For the limit case, when the gain A_0 becomes very high (~10), only the first interpolation stage determines the total offset. Since the first stage consists of $2^{N-N_{int}}-1$ stages with respect to 2^N stages in a flash A/D converter without interpolation, the maximum achievable improvement is equal to $\sim 2^{N_{int}}$, which is the result of equation 3.5 for large A_0. In practice, however, the A_0 is limited by the achievable linear region of the pre-amplifiers and the achievable gain-bandwidth product. The pre-amplifiers need to be linear for a sufficiently large input signal swing for proper interpolation.

From 3.5 we can conclude that already for moderate gain values ($A_0 \sim 4$) the improvement in intrinsic capacitance is close to $2^{N_{int}}$. This improvement is caused by reducing the number of required accurate pre-amplifiers and therefore reducing the intrinsic capacitance. When interpolation is used with only a small gain factor, for example for very high-speed converters [14], the total intrinsic capacitance is not reduced. However, interpolation is beneficial, since at the input fewer amplifiers with high accuracy are required, which reduces the input capacitance of the first stage of the converter.

Interpolation reduces (under the condition of sufficiently high A_0) the total intrinsic capacitance of the A/D converter, but it does not reduce the mismatch effects of each individual amplifier. The next paragraph describes a technique which improves the overall A/D converter linearity without increasing the size of the input transistors.

3.2.3 Averaging

Architecture

Averaging [17, 18, 19, 14] is a technique which improves the overall linearity of the A/D converter by averaging the error of an individual amplifier with several neighboring amplifiers. This is achieved by sharing the outputs of adjacent pre-amplifiers by inserting averaging resistors. The pre-amplifiers with averaging resistors and the effects of these averaging resistors are shown in figure 3.6 [19].

Ideally, the relationship between input and output is a straight line, but because of offset errors (indicated by the open dots in figure 3.6a), a random pattern is shown around the ideal curve as indicated by the thin lines connecting the open dots. The output resistors of the pre-amplifiers are indicated as the resistors between the open dots and the black dots. The averaging resistors are connected between the black dots. As can be seen, lowering this averaging resistor R_2 pulls the curve to a straight line, which reduces the pre-amplifier offset effect. Increasing the output resistances R_1 of the pre-amplifiers also increases this effect. In fact the amount

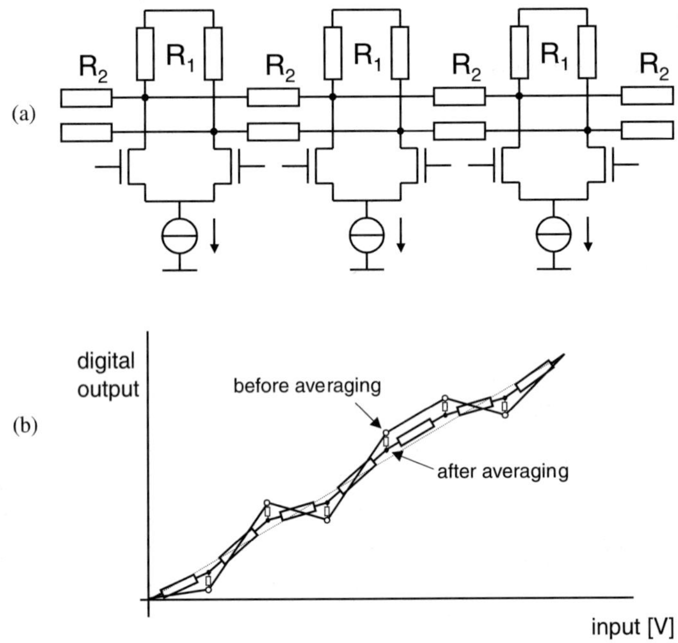

Figure 3.6: *Pre-amplifiers with averaging (a) and the effect of averaging on the transfer curve (b) [19]*

of averaging is a function of the ratio $\frac{R_2}{R_1}$. Figure 3.7 shows the reduction of the mismatch effect as a function of the ratio $\frac{R_2}{R_1}$ in the case of an infinite pre-amplifier array [14] and an infinite linear input range of the pre-amplifiers.

As can be concluded from figure 3.7, a smaller $\frac{R_2}{R_1}$ ratio reduces the effect of mismatch of the pre-amplifiers. The first assumption in figure 3.7 is that the number of pre-amplifiers is infinite. However, since the pre-amplifier array is not infinite in practice, the $\frac{R_2}{R_1}$ ratio cannot be reduced infinitely. This effect changes the position of the zero-crossings of the pre-amplifiers at both ends of the array. The positions of the zero-crossings can be restored by changing the reference voltages such that the zero-crossings after averaging are at their wanted position [14]. However, unequally spaced reference voltages are undesirable for matching. A more accurate compensation is to apply out-of-range circuitry, such that the zero-crossings of the in-range pre-amplifiers remain at their wanted position. By using an averaging termination circuit, which scales the resistors at both ends of the array, the required number of added over-range pre-amplifiers can be minimized. The operation of this termination circuit is described in [14]. The total number of out-of-range pre-amplifiers p required is a function of the ratio $\frac{R_2}{R_1}$ and is described by the following equation:

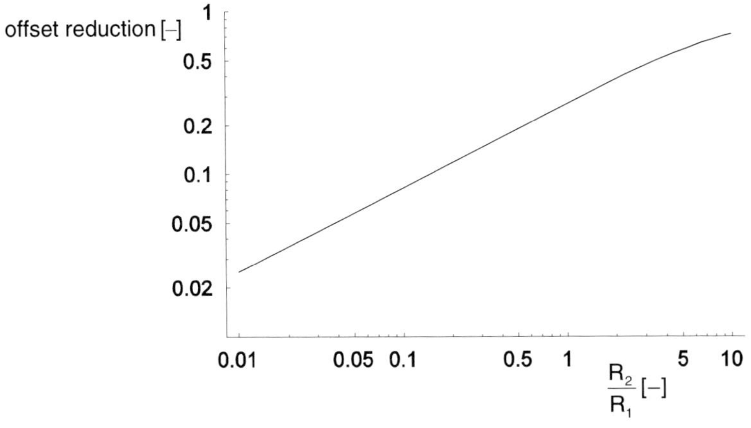

Figure 3.7: *Reduction in mismatch effect as a function of the ratio $\frac{R_2}{R_1}$, of an infinite pre-amplifier array and an infinite linear input range of the pre-amplifiers*

$$p = \sqrt{1 + 8\frac{R_1}{R_2}} - 1 \tag{3.6}$$

For high order of averaging, more out-of-range pre-amplifiers are therefore required, which increases the intrinsic capacitance.

The second assumption in figure 3.7 is the infinite linear range of the pre-amplifier input stages. Since in practical realizations of pre-amplifiers the linear input range is limited, the amount of offset reduction of a certain pre-amplifier is limited by the number of pre-amplifiers which are also in their linear input range. In general, the offset reduction after averaging is at maximum reduced by a factor $\sqrt{N_{linear}}$, where N_{linear} is the number of pre-amplifier stages operating in the linear input range [19]. This is illustrated in the example where if the averaging resistance R_2 is equal to zero, and for example 5 pre-amplifiers are in their linear input range, then these 5 pre-amplifiers can be seen as one lumped pre-amplifier, with a size that is 5 times larger. This reduces the mismatch with $\sqrt{5}$.

Intrinsic capacitance reduction

Since the offset of the pre-amplifiers is reduced due to averaging, smaller devices can be used in order to achieve the same accuracy. This reduces the intrinsic capacitance (the total input capacitance of the pre-amplifier array). The reduction in intrinsic capacitance is calculated for an array of $N_{preamps}$ (in-range) pre-amplifiers, which is, for example, the input stage of the comparators in a flash A/D converter. This pre-amplifier array is shown in figure 3.8.

Figure 3.8: *Pre-amplifier array, consisting of* $N_{preamps}$ *pre-amplifiers*

When averaging is applied to this array, which means that $R_2 < \infty$, p (equation 3.6) out-of-range pre-amplifiers have to be added in order to keep the zero-crossings at the wanted position. These additional pre-amplifiers have to be added to the total $N_{preamps}$ of the array. The resulting total intrinsic capacitance with respect to the total intrinsic capacitance of the reference pre-amplifier array can be expressed as a function of the number of pre-amplifiers ($N_{preamps}$), the required out-of-range pre-amplifiers p and the offset reduction. The latter two are both a function of R_1 and R_2.

$$\frac{C_{intr,averaged\ array}}{C_{intr,array\ (R_2\to\infty)}} = \frac{\left(N_{preamps} + p\left(\frac{R_2}{R_1}\right)\right)}{N_{preamps}} \cdot \left(offset\ red\left(\frac{R_2}{R_1}\right)\right)^2 \quad (3.7)$$

The offset reduction (*offset red*) reduces the capacitance quadratically due to the quadratic relation between the capacitance and the required accuracy (2.2.1). The reduction in intrinsic capacitance can be drawn as a function of the ratio $\frac{R_2}{R_1}$ with the number of pre-amplifiers $N_{preamps}$ as a parameter.

As shown in figure 3.9, the maximum improvement which can be achieved equals the number of pre-amplifiers in the in-range part. The improvement can never exceed this value since this is equal to the required size (and the respective intrinsic capacitance) of one amplifier for sufficient matching. In this limit case the offset of all the pre-amplifiers is averaged for each zero-crossing. The maximum reduction in intrinsic capacitance is therefore $\frac{1}{N_{preamps}}$.

When more pre-amplifiers ($N_{preamps}$) are used, the reduction in intrinsic capacitance is larger, since more pre-amplifiers in their linear region contribute to the averaging. For a certain value of $\frac{R_2}{R_1}$, for example 0.1, the reduction in capacitance is larger for larger $N_{preamps}$. The reason for this is that the additional out-of-range pre-amplifiers (p) are independent of $N_{preamps}$, but are relatively smaller for larger $N_{preamps}$. For a small $\frac{R_2}{R_1}$ ratio the number of required out-of-range pre-amplifiers becomes impractical since the reference range also has to increase. For example,

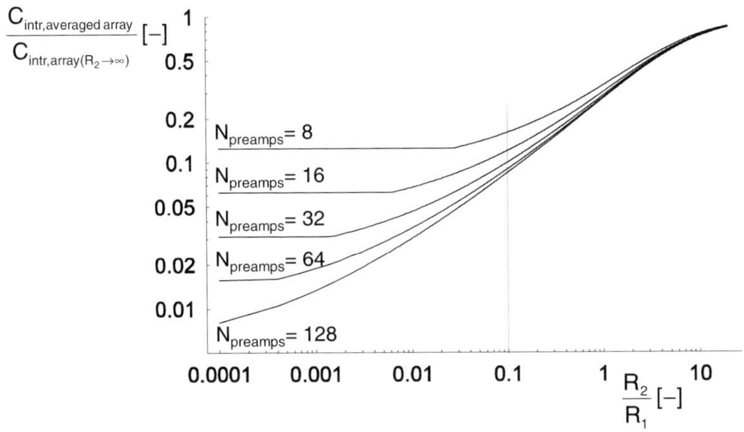

Figure 3.9: *Reduction in capacitance as a function of the ratio $\frac{R_2}{R_1}$, with the number of pre-amplifiers $N_{preamps}$ as a parameter*

with equation 3.6 it can be calculated that for $\frac{R_2}{R_1} = 0.1$ the number of additional pre-amplifiers is 8.

Due to its parallel signal processing, the flash A/D converter architecture is suitable for high-speed conversion. However, this parallel processing also limits the practical accuracy to 8 bits. Interpolation and averaging techniques reduce the intrinsic capacitance, thus reducing the power consumption. These techniques can also be combined [14].

3.3 Folding and interpolation

Architecture

For a higher number of bits it is obvious from figure 3.3 that the load of the comparators of a flash A/D converter becomes significantly large. To reduce the number of comparators and the number of input stages, the folding and interpolation architecture can be used [20, 21, 22, 16, 23, 24]. By folding the input signal, the same comparators can be used for several parts of the total range. The number of folds is called the folding factor F_f. Figure 3.10 shows one folding signal with a folding factor of 8.

However, the analog circuitry is not able to make the sharp transitions shown in figure 3.10. Nevertheless, between the transitions there is a linear region. To ensure linearity over the whole fine range a few parallel folding signals are

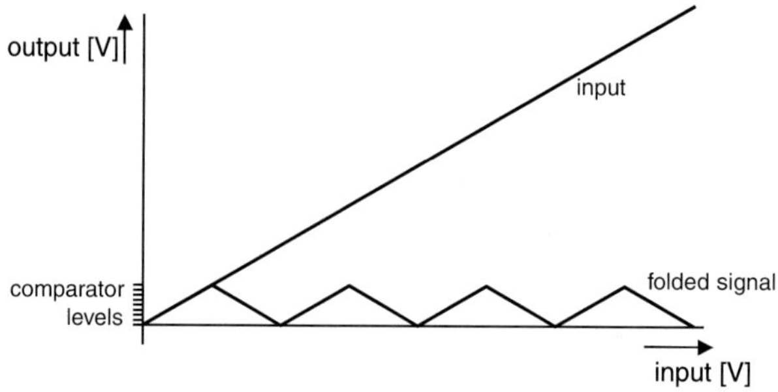

Figure 3.10: *Folding of input signal*

generated, of which only the 'linear' part is used for conversion. Figure 3.11 shows two parallel folding signals of which only the linear parts are applied to comparators.

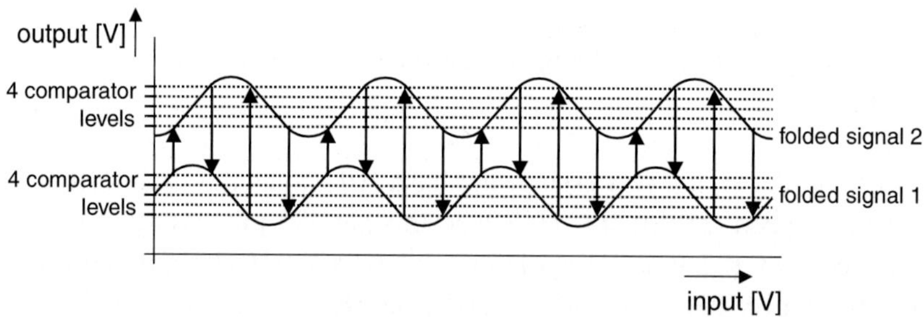

Figure 3.11: *Parallel folding by generating two folding signals*

Instead of taking one folding signal with 8 detection levels (figure 3.10) it is also possible to use 2 folding signals shifted with respect to the input signal with 4 detection levels (figure 3.11) each. This can be expanded to 8 folding signals with one detection level each. However, generating these folding signals requires the same amount of hardware as a full flash converter, since each zero-crossing is generated by one differential pair. To reduce the required number of differential pairs, interpolation can be used to generate additional detection levels (or zero-crossings). Between the folding signals shown in figure 3.11 additional detection levels are generated by an interpolating resistor ladder [20, 22], as shown in figure 3.12.

The number of additional interpolated levels is the interpolation factor (F_{int}). For example, 16 detection levels can be generated with 4 folding amplifiers and 4 times interpolation. The interpolation factor is limited by the requirement of integral non-linearity (INL). A larger interpolation factor requires a sufficiently high linear region of the folding-amplifiers [16]. This is in contradiction with

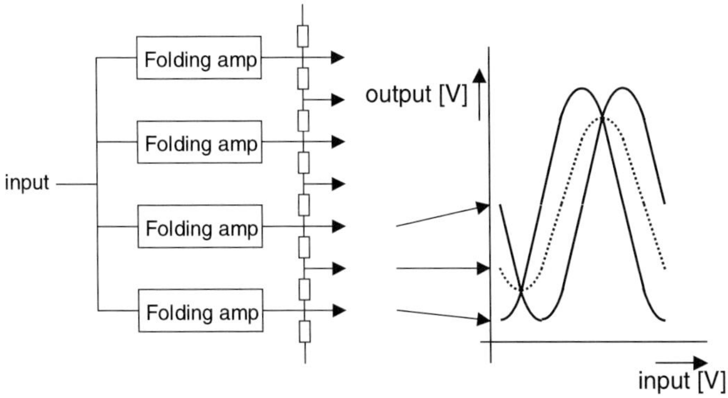

Figure 3.12: *Interpolation of folding signals*

proper folding which requires a small linear region, as will be explained later. An interpolation factor of 8 or 16 is often used [16, 19, 23]. For a proper conversion of the whole input range the different possible input voltages which correspond to the same folding output signal also have to be distinguished. This is done by connecting the input signal directly to additional comparators which generate the most-significant-bits (MSBs) of the total A/D converter. The block diagram of the total folding A/D architecture is shown in figure 3.13.

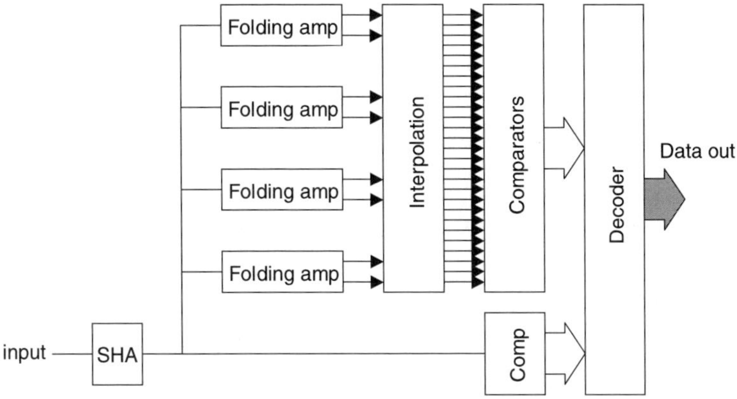

Figure 3.13: *The folding and interpolation architecture*

The sample-and-hold amplifier (SHA) at the input is required for high-frequency input signals because the folding pre-processing acts as an amplitude-dependent frequency multiplier [16]. Since the folding and interpolating A/D architecture makes its decision in one clock cycle, like the flash architecture it is very suitable for high-speed applications.

Accuracy

The input signal is sampled by the SHA. The noise sampled in and generated by this SHA directly influences the accuracy of the conversion. The accuracy of a folding architecture is mainly determined by the accuracy of the input transistors of the folding amplifiers. The offset of these input pairs determines the INL of the converter since this directly influences the positions of the zero-crossings. They must therefore have the accuracy of the overall A/D converter accuracy. When sufficient gain in the folding amplifiers is applied the comparator offset requirement is reduced by this gain. Other accuracy-determining factors are the size of the linear region of the folding amplifiers [16].

Intrinsic capacitance

When the input-referred offset of the folding amplifiers and the comparators is equally distributed, the total intrinsic capacitance of the folding amplifiers input stage is calculated in a similar way to the flash A/D architecture:

$$C_{intr,\ foldamps} = \frac{2^N}{F_{int}} \cdot C_{diff,input,matching} = \frac{C_{ox}^{\square} A_{V_T}^2}{3F_{int} V_{pp}^2} \cdot 2^{3N+8} \qquad (3.8)$$

where F_{int} is the interpolation factor which is applied to the A/D converter and the same yield conditions as in the flash architecture are considered. Because of folding, the number of required comparators is less than in the flash architecture and the gain in the folding amplifiers reduces the required comparator accuracy. The intrinsic capacitance from the comparators is calculated by:

$$C_{intr,\ foldcomps} = \frac{2^N}{F_f} \cdot C_{diff,input,matching} = \frac{C_{ox}^{\square} A_{V_T}^2}{3F_f {A_{fa}}^2 V_{pp}^2} \cdot 2^{3N+8} \qquad (3.9)$$

where F_f is the folding factor and A_{fa} is the gain of the folding amplifier. F_f gives the number of times a comparator is reused in the total signal range. A larger value means that fewer comparators are required, although the demand on the linear region of the folding amplifiers increases. A small linear region is required because otherwise more than one differential pair influences a zero-crossing at the output of the folding amplifier. This problem can be solved by using transistors in weak inversion to generate a small V_{gt}. To circumvent slow operation due to the large accumulated junction capacitances at the output of the folding amplifier a trans-resistance amplifier is used [22]. Another solution is given by using cascading of the folding action [19]. This relaxes the V_{gt} demands but the number of accurate input stages remains the same. The gain of the folding amplifiers A_{fa}

is limited by the limitation in linear region of the folding amplifiers, a large gain together with a large linear region will involve more power.

The total intrinsic capacitance of a folding A/D converter can be calculated by summing the results of equation 3.8 and 3.9:

$$C_{intr,\ folding} = C_{intr,\ foldamps} + C_{intr,\ foldcomps} \tag{3.10}$$

The result from equation 3.10 is shown in figure 3.14 as a function of N, with the folding factor F_f and the interpolation factor F_{int} as a parameter.

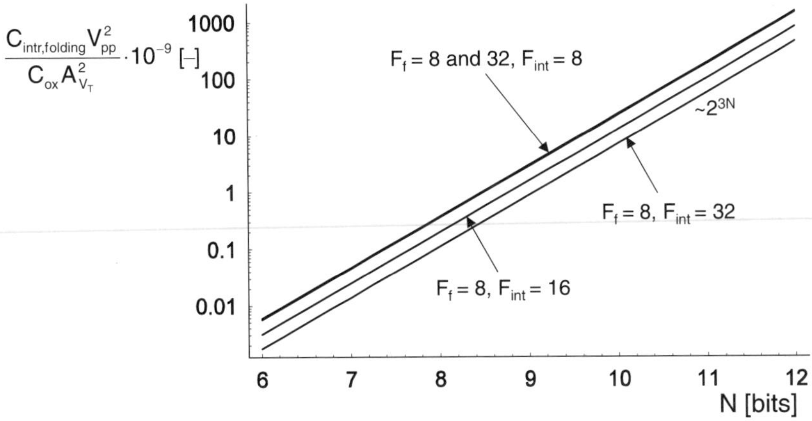

Figure 3.14: *Total intrinsic capacitance of a folding A/D converter as a function of the number of bits (N), with F_f and F_{int} as a parameter*

As can be seen in this figure, the folding factor does not influence the intrinsic capacitance. This is caused by the fact that in this calculation the folding amplifier gain A_{fa} is assumed to have a moderate value of only 4. The folding amplifier input transistors are then dominant in the overall capacitance and the number of comparators hardly influences the intrinsic capacitance. The interpolation factor is therefore of greater importance. But doubling the interpolation factor only reduces the number of input transistor pairs by half, which reduces the input capacitance of of the dominating folding amplifiers only by half. And also the linearity requirement increases for a larger interpolation factor. Generally, folding and interpolating A/D converter architectures reduce the total capacitive load with respect to a flash A/D converter only by the interpolation factor (8-16).

In the comparison of the A/D converter architectures it will be made clear that the intrinsic capacitance of the SHA in a folding A/D converter is much less than the intrinsic capacitance due to matching requirements. It is therefore ignored in the calculation of the total intrinsic capacitance of a folding A/D converter.

3.4 Two-step

By quantizing the analog signal in two steps, the number of comparators can be reduced significantly [25, 26]. A rough estimation determines in which part of the range the input signal lies and a more accurate quantization determines the digital code inside this part of the range. This section discusses the two-step architecture. A more general name is the subranging architecture, since more than two steps can also be used for quantization [27].

Architecture

A two-step A/D converter divides the quantization into two steps [26]. First a rough estimate is made with the coarse quantizer in the part of the range where the input signal is. This information is used to subtract the coarse signal with a D/A converter from the input signal. The remaining residue signal is then quantized in the fine quantizer to generate the LSBs. This is illustrated in figure 3.15.

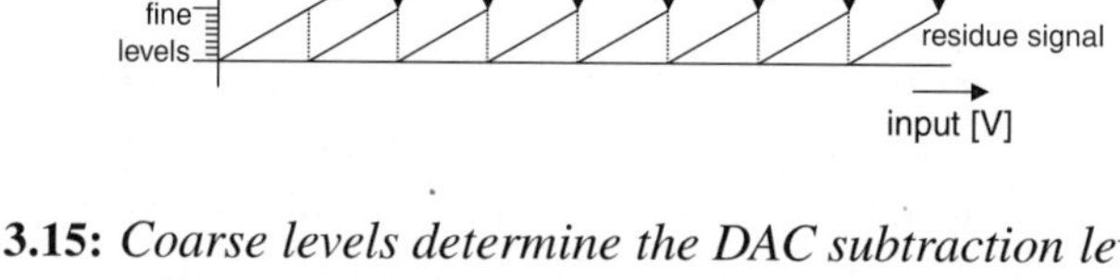

Figure 3.15: *Coarse levels determine the DAC subtraction levels to generate the residue signal*

Before the fine quantization is done, the residue signal is amplified by the residue amplifier to reduce the accuracy requirements in the fine A/D converter. Since the coarse and the fine quantization do not take place at the same time, the use of a sample-and-hold amplifier (SHA) is obligatory. The total two-step or subranging architecture is shown in figure 3.16.

To make a proper conversion, first a settling phase is required for the coarse conversion and, following this, the signal has to be amplified and applied to the fine

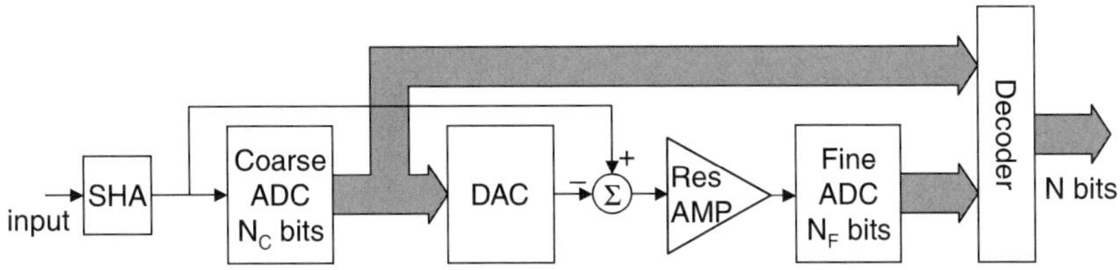

Figure 3.16: *Two-step A/D converter architecture*

converter. This therefore requires twice the settling time with respect to the flash and folding and interpolating architecture, although the settling requirements during coarse conversion are relaxed with respect to the settling during fine conversion, as will be explained later.

Accuracy

The first component that determines the accuracy of the two-step converter is the SHA. Noise generated and sampled in the SHA deteriorates the analog input signal which has to be quantized by the A/D converter. This generated and sampled noise has to be sufficiently low. To perform a proper total conversion the coarse A/D converter has to have an accuracy such that the resulting residue signal is always in the range of the fine A/D converter. This means that the comparators in the coarse A/D converter have to be accurate at the overall A/D converter accuracy, which is N bit. An error made in the coarse conversion results in an out-of-range residue signal [28]. The coarse decision drives a D/A converter, which generates an analog signal representing the coarse code. This D/A converter output is subtracted from the analog signal held by the SHA. It is obvious that the signal generated by the D/A converter has to be N-bit accurate. But this does not contribute to the power when the coarse references are used [28] and is therefore omitted from the intrinsic capacitance calculation. The resulting residue signal is applied to a residue amplifier. An offset in this residue amplifier results in shifting the residue signal within the fine range and causes the residue signal to be out-of-range in the fine A/D converter. The accuracy therefore needs to be the total A/D converter accuracy of N bit. The total offset error of the coarse A/D converter and the residue amplifier need to be within one LSB of the total A/D converter. This is shown in figure 3.17.

If the residue signal remains within the grey area of figure 3.17, the transfer of the total A/D converter is always monotonic. This means that the coarse decision and the offset of the residue amplifier always have to be within $\frac{LSB}{2}$. The accuracy requirements of the fine A/D converter are relaxed due to the gain of the residue

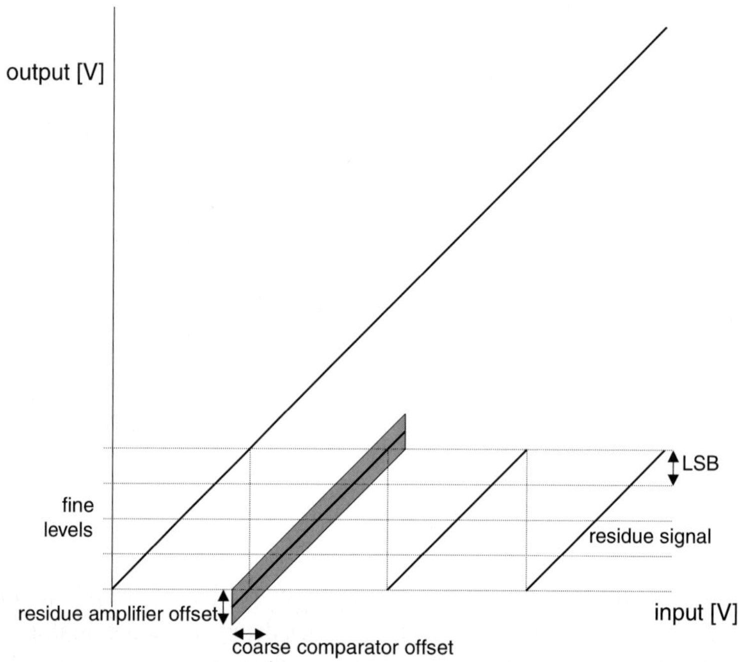

Figure 3.17: *Maximum allowed offset of both coarse A/D converter and residue amplifier, indicated by the grey area*

amplifier. The requirements of the fine comparators are the total A/D converter accuracy divided by the gain of the residue amplifier.

Redundancy

The accuracy requirement of the subrange selection results in high demands on the coarse A/D converter, as described above. When the coarse A/D converter makes an error, missing codes occur since the corresponding residue signal does not fit in the selected subrange. Figure 3.18a shows a proper subrange selection with its output code. Figure 3.18b shows the effects of an incorrect subrange selection.

As can be seen in figure 3.18b, the errors in the coarse A/D converter cause the residue signal to be out of range in the fine A/D converter. The coarse A/D converter accuracy requirements can be relaxed by using overlap of subsequent subranges [29]. This means that additional comparator zero-crossings are present outside the fine A/D converter range. The result of these additional zero crossings is that near a subrange transition a wanted code can originate from both subranges

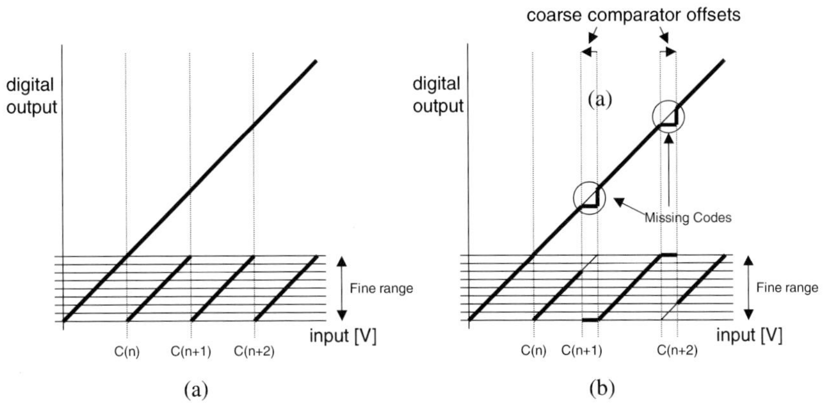

Figure 3.18: *Residue signals resulting from a proper subrange selection (a) and errors due to coarse comparator offset when no subrange overlap is used (b)*

around this subrange transition. Figure 3.19 shows the effect of coarse comparator offsets on the overall transfer of the A/D converter when subrange overlap is used.

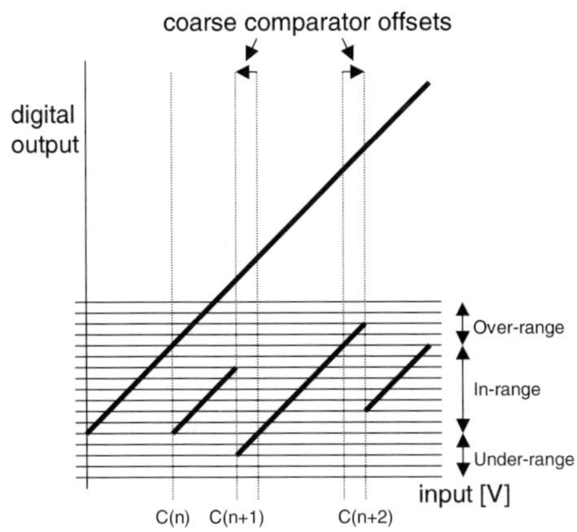

Figure 3.19: *No subrange transition errors when subrange overlap is used*

The digital encoding of the digital signals coming from the coarse and the fine A/D converter needs to resolve the errors from the coarse A/D converter. A schematic overview of this is shown in figure 3.20.

The encoder needs to detect whether over- or under-range has occurred and one coarse LSB is then subtracted or added to the coarse digital code accordingly.

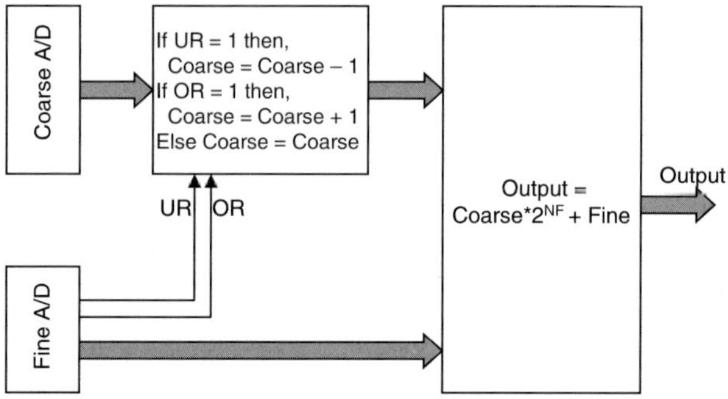

Figure 3.20: *Digital decoding of over- and under-range occurrences*

Redundancy[1] is a very powerful optimization with respect to required accuracy. For example, when the fine range is doubled, the accuracy requirements of the coarse A/D converter are reduced to half an LSB of this coarse converter, e.g. a 10-bit A/D converter with 5 coarse and 5 fine bits. Using half an additional subrange on both sides increases the intrinsic capacitance of the fine A/D converter by 2, but reduces the load capacitance of the coarse A/D converter by a factor of 2^{12}!

Adding over-range to the fine A/D converter not only reduces the coarse A/D accuracy, but also means the offset requirements of the residue amplifier can be reduced as well. Since residue amplifier offset causes a shift of the residue signal within the fine A/D converter range, this error can be traded-off with the error in the subrange selection. Redundancy therefore reduces both A/D converter accuracy and the residue amplifier offset requirement.

Generally, the accuracy of the fine A/D converter is increased by one bit. Compared to the bare minimum, a larger fine accuracy does not further reduce the requirements of the coarse A/D converter since then the monotonicity of the coarse A/D converter is not guaranteed.

Intrinsic capacitance

The coarse A/D converter accuracy and the residue amplifier accuracy are linked together. If it is assumed that the over-range of the fine A/D converter is one bit, the sum of the error of the coarse A/D converter and the residue amplifier offset:

[1] Redundancy can be used in all types of converters where a selection has to be made for the following step. These are the pipe-lined, two-step and successive approximation (SAR) architectures, as explained later.

$$\begin{aligned} V_{offset,tot} &= V_{offset,res\ amp} + V_{offset,crs\ comp} \\ &= V_{offset,res\ amp} + x \cdot V_{offset,res\ amp}, \end{aligned} \tag{3.11}$$

with $x = \frac{V_{offset,crs\ comp}}{V_{offset,res\ amp}}$, is allowed to be half an LSB of the coarse A/D converter resolution. In this case the over-range of the fine A/D converter is able to correct these errors. With the use of equation 2.6 the total offset is expressed in the capacitance of the input of the residue amplifier stage:

$$V_{offset,tot} \propto \frac{1}{\sqrt{C_{res\ amp}}} + \frac{x}{\sqrt{C_{res\ amp}}} \tag{3.12}$$

When equation 3.12 is rewritten, the total capacitance of both the coarse A/D converter and the residue amplifier stage can be expressed in x, $V_{offset,tot}$ and the number of coarse bits N_C:

$$\begin{aligned} C_{total} &= C_{res\ amp} + 2^{N_C} \cdot C_{crs\ comp} \\ &\propto \sqrt{\frac{x+1}{V_{offset,tot}}} + 2^{N_C} \cdot \frac{1}{x^2}\sqrt{\frac{x+1}{V_{offset,tot}}} \end{aligned} \tag{3.13}$$

By minimizing equation 3.13, the optimum value for the ratio x between the offset of the coarse comparators and the residue amplifier results in:

$$x = \frac{V_{offset,crs\ comp}}{V_{offset,res\ amp}} = \frac{1}{2^{\frac{N_C}{3}}} \tag{3.14}$$

In the derivation above it is assumed that the coarse A/D converter consists of a full flash converter without enhancement techniques as interpolation or averaging. It is obvious that a different coarse A/D converter influences this ratio. Using the ratio of equation 3.14, the intrinsic capacitance of the residue amplifier can be expressed as a function of the number of coarse bits N_C:

$$C_{intr,res\ amp} = \left(1 + \sqrt{2^{\frac{2N_C}{3}}}\right)^2 \cdot \frac{C_{ox}^{\square} A_{V_T}^2}{3V_{pp}^2} \cdot 2^{2N_C+8} \tag{3.15}$$

The part after the brackets is the intrinsic capacitance required for a maximum of $\frac{LSB}{4}$ offset error. The part between the brackets is the result of an optimum distribution of the offset errors over the coarse conversion and the residue amplifier (see equation 3.14). The intrinsic capacitance of the coarse A/D converter is calculated in a similar way to the flash architecture (equation 3.2). With the only difference that the accuracy of the coarse decision is dependent on the ratio of equation 3.14:

$$C_{intr,crs\ ADC} = \frac{\left(1+\sqrt{2^{\frac{2N_C}{3}}}\right)^2}{2^{\frac{2N_C}{3}}} \cdot \frac{C_{ox}^{\square} A_{V_T}^2}{3V_{pp}^2} \cdot 2^{3N_C+8} \tag{3.16}$$

The accuracy of the fine A/D converter is chosen $\frac{LSB}{4}$ (appendix A), since the fine conversion determines the positions of the quantization levels contrary to the coarse conversion, which only selects the proper subrange. The offset requirement is divided by the gain of the residue amplifier, which is in general equal to half of the number of levels of the coarse A/D converter (2^{N_C-1}) since the signal applied to the fine A/D converter consists of two subranges due to redundancy. The intrinsic capacitance of the fine A/D converter is then:

$$C_{intr,fine\ ADC} = \frac{C_{ox}^{\square} A_{V_T}^2}{3V_{pp}^2} \cdot 2^{3N-3N_C+12} \tag{3.17}$$

$C_{intr,res\ amp}$, $C_{intr,crs\ ADC}$ and $C_{intr,fine\ ADC}$ are dependent on N_C. The optimum value of N_C can be calculated by minimizing the sum of these capacitances. The optimum value of N_C can be approximated by:

$$N_{C,opt} \approx \frac{N+1}{2} \tag{3.18}$$

The total intrinsic capacitance resulting from the matching requirements in a two step A/D converter is the sum of equation 3.15, 3.16 and 3.17:

$$C_{intr\ match,twostep} = C_{intr,res\ amp} + C_{intr,crs\ ADC} + C_{intr,fine\ ADC} \tag{3.19}$$

Figure 3.21 shows the total capacitance of the two-step architecture and the contributions separately with optimum value for N_C (equation 3.18), as a function of N.

In figure 3.21 the coarse and the fine A/D converters are based on the full flash A/D converter without enhancement techniques. If these techniques, such as interpolation or averaging, are applied to these stages, the total capacitance is reduced significantly, since the intrinsic capacitance from the residue amplifier, which is not reduced by such enhancement techniques, is not dominant in the total intrinsic matching capacitance.

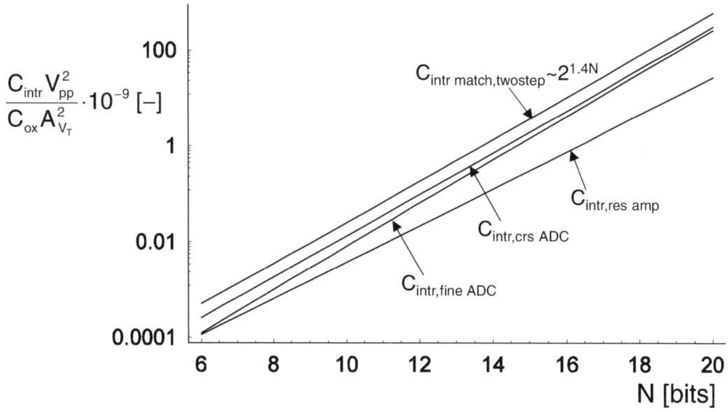

Figure 3.21: *Total intrinsic matching capacitance of a two-step A/D converter as a function of the number of bits (N) and the separate contributions for optimum N_C*

The SHA which is required at the input of the two-step architecture (figure 3.16 comprises a capacitor to hold the analog signal during the hold phase. During sampling, in addition to the input signal thermal noise is also sampled onto this capacitance. The amount of thermal noise is calculated by equation 2.12. This sampled thermal noise has to be sufficiently low, as explained in section 2.2.2, to achieve an accuracy of N bits. When a buffer drives this capacitance (section 2.2.2), the required hold capacitance is given by:

$$C_{intr,SHA} = \frac{8 \cdot 2^{2N+2} NEF\, kT}{V_{pp}^2}, \tag{3.20}$$

with *NEF* as the noise excess factor [7]. Equation 3.20 can be drawn as a function of the number of bits N. This is shown in figure 3.22.

The sum of the values derived from figure 3.21 and figure 3.22 is the total intrinsic capacitance of a two-step A/D converter. The intrinsic capacitance originating from the sample and hold capacitance is small with respect to the intrinsic matching capacitance. For example, in 0.18 μm CMOS, with $A_{V_T} = 5$ mV, $C_{ox}^{\beta} = 7$ fFμm^2, $V_{pp} = 1$ V, $NEF = 4$ and $N = 12$, the total intrinsic matching capacitance is 35 pF, while $C_{intr,SHA}$ is 13 pF. The coarse and fine A/D converters are dominant in the total capacitance for accuracies up to 12 bits. This capacitance can be reasonably reduced by architectural enhancements. However, when both the coarse and fine flash A/D converters are optimized by using the techniques described in section 3.2, this difference becomes smaller. For higher accuracies the intrinsic capacitance from the SHA can even become dominant.

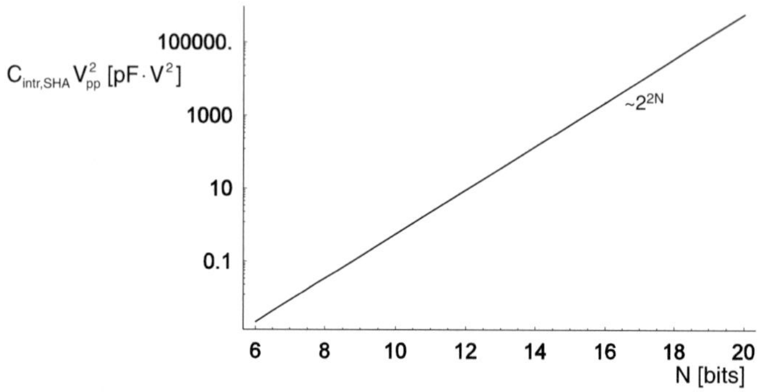

Figure 3.22: *Total intrinsic capacitance of a SHA in a two-step A/D converter as a function of the number of bits (N), with NEF* $= 4$

The two-step architecture provides a large reduction in intrinsic capacitance compared to the flash architecture. However, to make a proper conversion, it is very important in a two-step structure that the fine range fits exactly in a coarse range. In this derivation above it is assumed that the circuitry that provides this guarantee has a negligible contribution to the intrinsic capacitance. In the next chapter of this book architectural changes are presented to deal with these range and gain topics. In the two-step architecture the coarse and the fine quantization take place within one clock period. The next section describes the pipe-line architecture where the quantization operations take place serially.

3.5 Pipe-line

Architecture

The pipe-line architecture [29] shows a large similarity to the two-step architecture. A two-step A/D converter consists of 2 stages, while a pipe-line A/D converter generally consists of N stages, each including a SHA, an A/D converter, a D/A converter, a subtractor and an amplifier. The difference between the two-step and the pipe-line architecture is the SHA after each stage. The conceptual one-bit-per-stage pipe-line is shown in figure 3.23.

The basic operation is the same as in the two-step architecture. The first stage samples the analog input signal. This signal is converted to the digital domain to estimate the held input. The result from this quantization is applied to the D/A converter, which generates an analog signal that represents the digital A/D

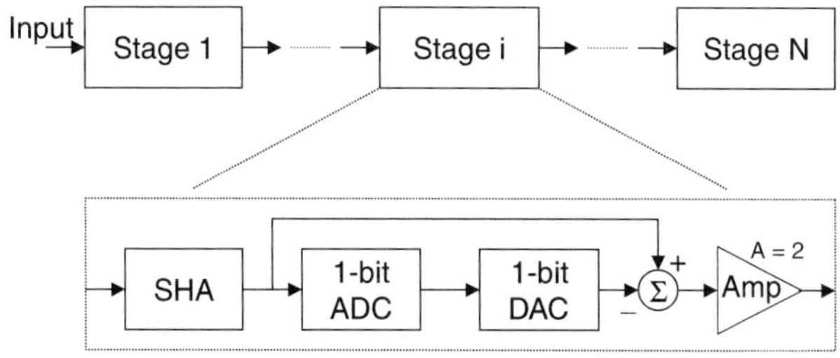

Figure 3.23: *Pipe-line A/D converter architecture with one bit per stage*

converter output. The D/A converter output is subtracted from the held analog input signal. This produces the residue signal, which is amplified before it is applied to the next stage. Contrary to the two-step architecture the second stage then samples this amplified residue signal and the same operations take place. This is continued until all stages have performed their conversion. All digital output values are combined to generate the digital output code. Due to the pipe-lining concept (caused by the inserted SHAs) the first stage can start sampling a new analog input signal while the second stage processes the previous sample. The speed of the conversion is therefore determined by the speed of a single stage. The drawback of this, however, is a larger latency. Figure 3.23 shows the case for one bit per stage, although more configurations are possible, such as one and a half bits per stage to resolve A/D converter errors [29] or a first stage with three bits and the remaining stages with one and a half bits per stage [30]. First the one and a half bit first stage is considered and then the advantages with respect to a decreasing total intrinsic capacitance are described.

One and a half bits per stage relaxes the demands on the comparators (inside the ADC in figure 3.23) since it adds redundancy, but does not change the requirements for the capacitors in the SHA with respect to noise or absolute residue gain requirements. Since the capacitors in the SHAs are dominant in the total intrinsic capacitance, for the sake of simplicity the conceptual one bit per stage pipe-line converter is considered, while the comparator requirements are left out of the discussion.

During a conversion cycle, first the signal is sampled and a conversion is performed. Subsequent to that the DAC has to be set with the proper code and the subtraction has to be carried out. Within one clock cycle there are therefore two settling actions that have to be performed, as in the two-step architecture.

Accuracy

The accuracy of the pipe-line architecture is determined mainly by two factors. The noise sampled in each pipe-line stage reduces the achievable SNR of the converter. The one (and one and a half) bit per stage pipe-line architectures rely on the exact factor of two gain in each stage. This also puts constraints on the matching of capacitors. First the origin of noise is discussed.

In each stage of the pipe-line converter the analog signal is re-sampled. This means that in each stage all of the wide-band noise generated in the switches and buffers of that stage is folded back into the sampling base-band. Due to the gain after each pipe-line stage, the contribution to the total noise of the last stages is less than the contribution of the first stages (referred to the input). Since the noise power generated in each stage is inversely proportional to the load capacitance (equation 2.8), the required capacitances can be scaled. For example, the scaling factor can be chosen such that the contribution in noise is equal for all stages. This, however, causes the power to be dominated by the first stage [31]. For the conceptual pipe-line A/D converter (without parasitic capacitances) the optimum scaling factor is two for the subsequent stages [31]. This means that the first stage contributes the same amount of noise as the sum of all the remaining stages.

The gain factor of two is critical in the performance of the pipe-line architecture. This is equal for both the one bit or one and a half bit per stage architecture since redundancy does not solve gain errors. Generally, this gain factor is determined by the ratio of two capacitors. Due to capacitor mismatch [32, 33] this gain factor can deviate from its ideal value, which causes INL errors. The capacitor mismatch is area dependent. Larger capacitors have better matching. The total intrinsic capacitance is determined by the largest capacitance determined by noise or matching.

For the one bit per stage A/D converter shown in figure 3.23 decision-accurate comparators are required. However, when redundancy is applied (as in the two-step converter) the accuracy requirements of the comparators are reduced. Since the sampling capacitors are dominant in the total intrinsic capacitance, the comparators are left out of the calculation.

Intrinsic capacitance

In this section first the minimum required capacitance for noise considerations is calculated, then the matching of capacitors is discussed.

Due to re-sampling in each stage, the noise requirement can be considered separately for each stage. In this derivation it is assumed that for each stage there is one sampling action [34]. When a pipe-line scheme is chosen with two sample actions per stage, the required capacitance is two times larger. The contribution to the total noise of each stage is chosen such that the capacitance of a certain stage has the same value as the sum of the capacitances of all subsequent stages [31]. In this case the first stage produces half of the total noise. When a buffer drives this capacitance (section 2.2.2), the total required capacitance in this case can be calculated by (equation 2.8):

$$C_{intr,pipeline,noise,1} = 2 \cdot \frac{8 \cdot 2^{2N+2} NEF\, kT}{V_{pp}^2} \tag{3.21}$$

For the remaining stages the capacitance is scaled down by a factor of two. The required capacitance per stage as a function of the number of bits is then given by:

$$C_{intr,pipeline,noise,i} = \left(\frac{2}{2^{i-1}}\right) \frac{8 \cdot 2^{2N+2} NEF\, kT}{V_{pp}^2} \tag{3.22}$$

with i as the stage number ($1 \leq i \leq N$). For an N-bit pipe-line A/D converter the total intrinsic capacitance required for noise considerations can be calculated by summing the capacitances of all the stages:

$$C_{intr,pipeline,noise} = \sum_{i=1}^{N} C_{noise,pipeline,i} = \frac{2^{N+7}\left(2^N - 1\right) NEF\, kT}{V_{pp}^2} \tag{3.23}$$

The total intrinsic capacitance (equation 3.23) can be drawn as a function of the number of bits N. This is shown in figure 3.24.

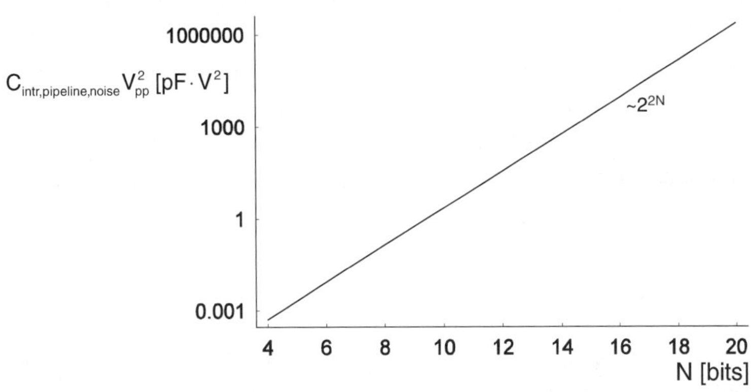

Figure 3.24: *Total capacitive load of a pipe-line A/D converter as a function of the number of bits (N), with NEF = 4*

In this figure *NEF* from equation 3.23 is assumed to be four. As can be seen in figure 3.24, the capacitive load grows exponentially with the required accuracy.

To calculate the matching requirements of the capacitors it is assumed that the following relation exists between the error due to mismatch in capacitance and the value of the capacitance [32, 33]:

$$\frac{\sigma_C}{C} = \frac{A_C}{\sqrt{C}} \tag{3.24}$$

This equation supposes that an error in the capacitance is caused by fluctuations in the thickness of the inter-metal layer. A_C is a technology constant, dependent on the quality of the processing of the inter-metal layer. For example, in a double poly process the inter-poly oxide thickness is well controlled and therefore A_C is generally smaller than in the case of back-end capacitors. A general pipe-line stage is shown in figure 3.25.

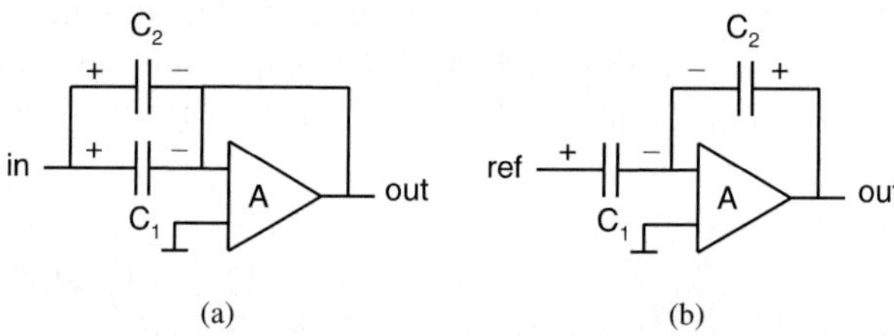

Figure 3.25: *A general pipe-line stage during sample phase (a) and hold and gain phase (b)*

In the case of one bit per stage, the signal is sampled on capacitors C_1 and C_2 during the sample phase. During the hold and gain phase the charge of C_1 is transferred to C_2, which is then put in feedback. To ensure a gain of two, both capacitors shown in figure 3.25 have to have a ratio of $\frac{C_1}{C_2} = 1$. The maximum error in the first stage, which results in a DNL (differential-non-linearity) error due to a mismatch between C_1 and C_2, normalized to an LSB is given by:

$$DNL = \frac{\Delta C_1 \cdot 2^{N-1}}{C_1} \tag{3.25}$$

with $\Delta C_1 = 4\sigma_{C_1} = 4A_C \cdot \sqrt{C_1}$. Rewriting equation 3.25 gives the minimum required input capacitance of the first stage for sufficient matching (equation 3.26).

$$C_{intr,pipeline,match,1} = C_1 + C_2 = 2 \cdot 2^{2N+6} A_C^2 \tag{3.26}$$

The maximum allowed DNL is set at $\frac{LSB}{4}$. For sufficient yield, the maximum allowed capacitance error has to be within $\frac{\sigma_C}{4}$. As with the total capacitance calculated for the noise limitation, the same rule of scaling can be applied for calculating the intrinsic capacitance for matching of the complete pipe-line converter. This gives the following result for the total intrinsic capacitance for matching requirements:

$$C_{intr,pipeline,match} = 2^{N+8}\left(2^N - 1\right)A_C^2 \tag{3.27}$$

The ratio between $C_{intr,pipeline,match}$ and $C_{intr,pipeline,noise}$ determines whether the intrinsic capacitance is determined by the noise or by the matching of capacitors. This ratio is given by:

$$\frac{C_{intr,pipeline,match}}{C_{intr,pipeline,noise}} = \frac{2A_C^2 V_{pp}^2}{kT \cdot NEF} \tag{3.28}$$

It can be observed that this ratio is independent of the resolution (N). This means that the ratio holds for all of the stages inside a pipe-line converter. When the ratio from equation 3.28 is smaller than 1, the intrinsic capacitance is determined by the noise requirements and is calculated with equation 3.23. A ratio larger than 1 means that the intrinsic capacitance is determined by the matching of the capacitors and is calculated with equation 3.27 or calibration is required. This is described in the next chapter. Figure 3.26 shows the ratio as a function of the capacitance matching parameter of a certain technology. This ratio is dependent on the signal amplitude and the *NEF*.

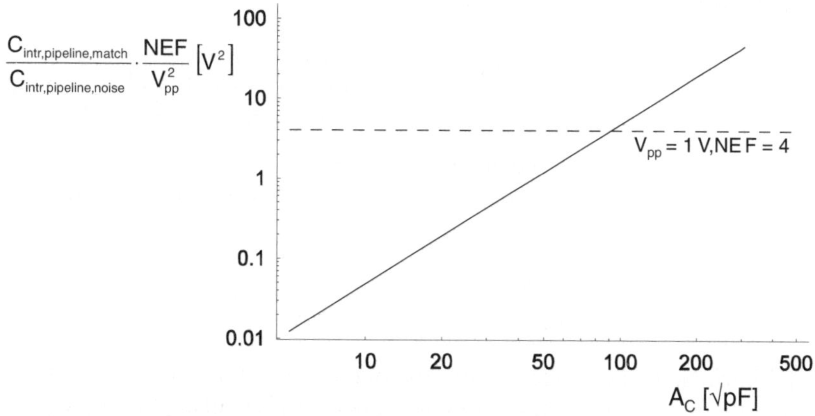

Figure 3.26: *The ratio of capacitor matching and noise requirements as function of the capacitance matching parameter* A_C

From figure 3.26 it can be read that for $V_{pp} = 1$ V and a noise excess factor of $NEF = 4$, the required capacitor matching is equal to $A_C = 91$ pF$^{\frac{1}{2}}$. When for a certain technology this matching is not achieved, the capacitors have to be scaled up for sufficient matching or calibration can be used to calibrate the gain factor of the respective stages. In [33] it is shown that when using a multi-bit first stage in a pipe-line A/D converter, the capacitor matching of the technology used is sufficient and no increased capacitance or calibration is required.

Multi-bit first pipe-line stage

Using a multi-bit first pipe-line stage reduces the requirements and the intrinsic capacitance of the subsequent stages [30, 35, 33, 36, 37]. As is shown above and in [31], the optimum scaling factor for the capacitances of succeeding stages is equal to the gain of the preceding stages. In the case of one bit (or one-and-a-half bits) per stage this factor is two. The total thermal noise, calculated with equation 3.21 and equation 3.22, is then:

$$P_{noise,tot,onebit} = \frac{kT}{C_{intr,pipeline,noise,1}} \cdot \left(2 - \frac{1}{2^{N-1}}\right) \tag{3.29}$$

When the pipe-line is implemented with a multi-bit first stage, the capacitor of the next stage can be scaled with the gain of the first stage, which is generally 2^{N_C}, where N_C is the first stage resolution. The next stages are again scaled with a factor of two. This means that the total noise contribution of all the stages after the first stage is reduced by a factor of two for each additional bit in the first stage. The total thermal noise is then given by:

$$P_{noise,tot,mltbit} = \frac{kT}{C_{intr,mltbitpipeline,noise,1}} \cdot \left(1 - \frac{1}{2^{N-2}} - \frac{1}{2^{N_C-1}}\right) \tag{3.30}$$

For equal thermal noise, this means that the ratio between the intrinsic capacitance of the first stage of a multi-bit pipe-line and a one-bit pipe-line is equal to:

$$\frac{C_{intr,mltbitpipeline,noise,1}}{C_{intr,pipeline,noise,1}} = \frac{1 - \frac{1}{2^{N-2}} - \frac{1}{2^{N_C-1}}}{2 - \frac{1}{2^{N-1}}} \tag{3.31}$$

The total intrinsic capacitance of both the one-bit-per-stage and the multi-bit first stage pipe-line can be expressed in their respective first stage capacitances:

$$C_{intr,pipeline,noise} = \left(2 - \frac{1}{2^{N-1}}\right) \cdot C_{intr,pipeline,noise,1} \tag{3.32}$$

$$C_{intr,mltbitpipeline,noise} = \left(1 - \frac{1}{2^{N-2}} - \frac{1}{2^{N_C-1}}\right) \cdot C_{intr,mltbitpipeline,noise,1} \quad (3.33)$$

The reduction in intrinsic capacitance of a pipe-line A/D converter when using a multi-bit first stage can now be calculated by dividing equation 3.33 by equation 3.32 and using equation 3.31:

$$\begin{aligned}\frac{C_{intr,mltbitpipeline,noise}}{C_{intr,pipeline,noise}} &= \left(1 - \frac{1}{2^{N-2}} - \frac{1}{2^{N_C-1}}\right)^2 \cdot \left(2 - \frac{1}{2^{N-1}}\right) \\ &\approx 2\left(1 - \frac{1}{2^{N_C-1}}\right)^2, \qquad (3.34)\end{aligned}$$

for N is larger than 8 bits. The reduction ratio from equation 3.34 is drawn as a function of the number of bits in the first stage. This is shown in figure 3.27.

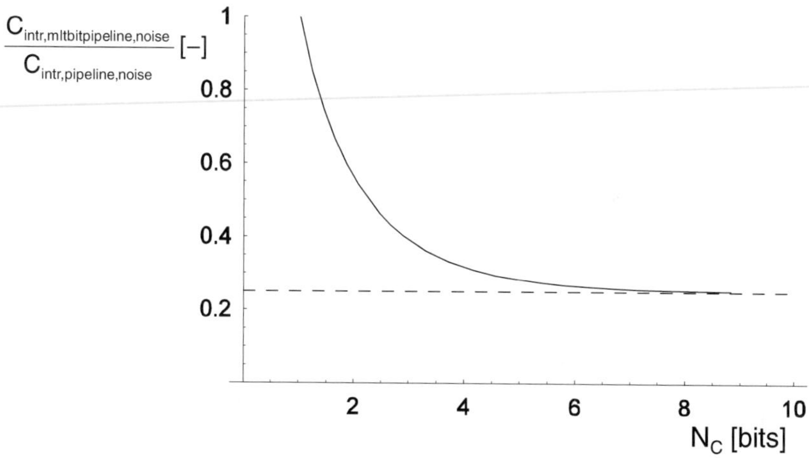

Figure 3.27: *The reduction ratio of intrinsic capacitance when using a multi-bit first stage in a pipe-line A/D converter as a function of the number of bits in the first stage*

As can be seen in figure 3.27, the maximum reduction ratio is 0.25. The intrinsic capacitance then equals the minimum required capacitance to sample the signal on a single sample-and-hold stage with an accuracy required for N bit. The capacitance matching requirements when using a multi-bit first stage are relaxed by a factor of $\sqrt{N_C}$, as has been demonstrated in [33]. In practice a value of 3 to 4 is used [30, 35, 33, 36, 37]. Using more bits in the first stage requires a more accurate flash A/D converter stage, but this does involve greater power consumption.

3.6 Successive approximation

The pipe-line architecture uses a separate stage for each bit (in the one-bit-per-stage configuration). The successive approximation architecture uses the same quantization stage for each bit [38, 39, 40].

Architecture

The successive approximation A/D converter converts the analog signal to a digital code by a binary search. The SAR converter, shown in figure 3.28, consists of a SHA, which has to hold the signal during the binary search.

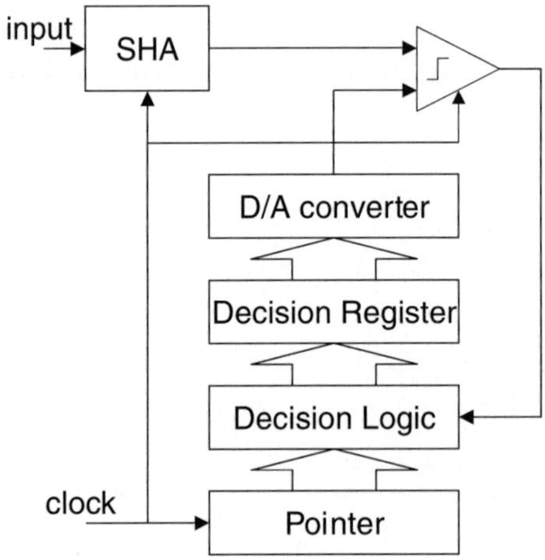

Figure 3.28: *Successive approximation A/D converter architecture*

At the start of this binary search, the MSB of the D/A converter is set to one. All the other bits are set to zero. The (only) comparator decides whether the input signal is larger or smaller than the signal from the D/A converter. After this decision, the MSB of the D/A converter is set according to this decision and the MSB − 1 is set to one. Again the comparator compares the input signal to this new D/A converter value. This binary search continues until all bits have the proper values. A conversion cycle is illustrated in figure 3.29.

The subtraction D/A converter can consist of a reference ladder with switches to select the reference closest to the input signal level. However, for large resolution (>8 bits) a large amount of switches is required. The subtraction D/A converter can also be based on charge redistribution [38, 40], which is done with capacitors. This, however, requires sufficient matching of capacitors.

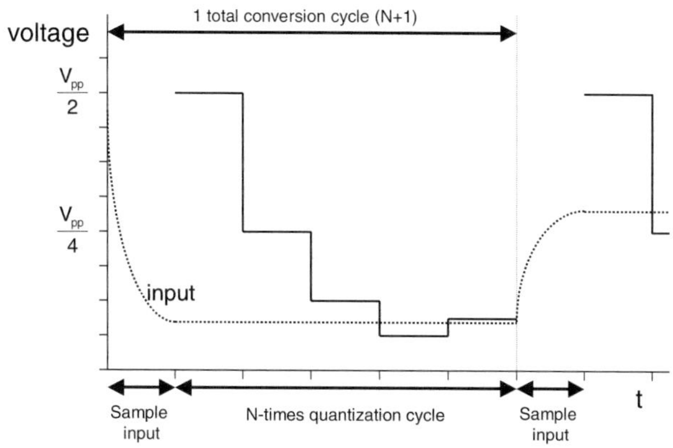

Figure 3.29: *Binary search of the output code*

As can be seen in figure 3.29, a full conversion cycle takes an input sample time plus N times a comparator decision cycle. In general this means that the conversion cycles of a SAR A/D converter has to be $N + 1$ times faster than the flash architecture to achieve the same Nyquist frequency. During the conversion cycles redundancy can be applied, which relaxes both the accuracy and timing demands of the quantization. The major power-consuming component in the SAR architecture is therefore generally the SHA. To optimize the power consumption a larger part than $\frac{1}{N+1}$ of the total sample period can be used for sampling the analog input signal. This means that the input sample time and the quantization cycles in figure 3.29 are not equally distributed over the full conversion cycle time. For the sake of simplicity, in the remainder of the calculation equal distribution (as shown in figure 3.29) is assumed.

Accuracy

As with the pipe-line architecture, both noise and capacitor matching are important for the accuracy. The noise generated (or sampled) in the SHA needs to be sufficiently low for the required accuracy, which is similar to the SHA of the two-step architecture. The D/A converter generally uses capacitors [38] (the same capacitors as in the SHA [40]) to generate the proper subtraction signals. The matching between the capacitors used in this subtractor determines the achievable accuracy of the SAR converter. The comparator accuracy is not important, since a comparator offset does not contribute to a conversion error. It only adds to the offset of the overall SAR converter.

Intrinsic capacitance

The capacitance in the SHA has to fulfill the noise requirements for a given number of bits N. This is calculated with the same formula as for the two-step converter (equation 3.20). However, if the matching of capacitors of the technology used is not sufficient for the given technology, the intrinsic capacitance has to be increased and is given by equation 3.26. Equation 3.28 can be used to verify the matching of the capacitors used.

The intrinsic capacitance of the SAR architecture is therefore very similar to that of the pipe-line converter. The difference is that the intrinsic capacitance of the SAR architecture is a factor of two less because there is only one stage. The intrinsic capacitance determined by the noise requirement as a function of the resolution (N) is shown in figure 3.30.

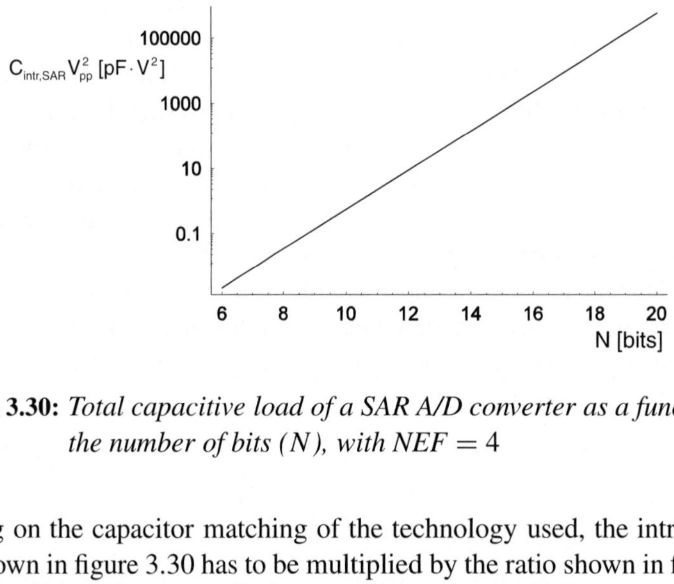

Figure 3.30: *Total capacitive load of a SAR A/D converter as a function of the number of bits (N), with $NEF = 4$*

Depending on the capacitor matching of the technology used, the intrinsic capacitance shown in figure 3.30 has to be multiplied by the ratio shown in figure 3.26.

3.7 Theoretical power consumption comparison

In the previous section the theoretical minimum total capacitive load has been derived for several architectures. These results can be used to compare the efficiency of the architectures as a function of the required number of bits N and the sample rate.

3.7.1 Figure-of-Merit (*FoM*)

The *FoM* is calculated by dividing the power consumption of the A/D converter by the achieved accuracy and the minimum of the sample frequency and two times the resolution bandwidth (ERWB). This is explained with the aid of figure 3.31 and given in equation 3.35.

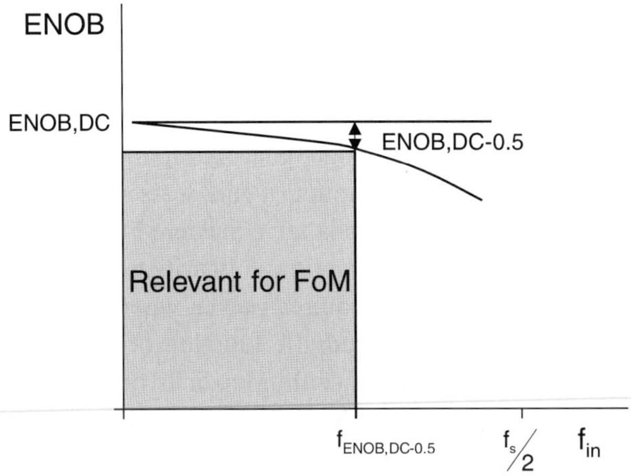

Figure 3.31: *Representation of the figure of merit (FoM)*

$$FoM = \frac{P}{2^{ENOB,DC} MIN(f_s, 2f_{ENOB,DC-0.5})} \qquad (3.35)$$

The *FoM* is a measure for the required power per achieved resolution per conversion. Today, the state-of-the-art *FoM* for Nyquist A/D converters is around 1 pJ/conversion, which means that 1 pJ per conversion cycle is required to convert an analog signal to a single bit.

3.7.2 Architecture comparison as a function of the resolution

Theoretical A/D converter comparison

Equation 3.35 is used for an objective comparison between the architectures described earlier. Since the power consumption is proportional to the intrinsic capacitance 2.15, in the calculation the calculated intrinsic capacitance for each architecture is used as a measure for the power P in equation 3.35. The intrinsic capacitance determined by matching and noise is dependent on different technology parameters. To be able to make a comparison between the architectures, realistic values for these technology parameters have to be chosen. To compare the figure-of-merits therefore, as an example the technology parameters are set at values for a 0.18 μm technology. Those values are: $V_{pp} = 1$ V, $C_{ox}^{\square} = 7$ fF/μm^2, $A_{V_T} = 5$ mV and $NEF = 4$. However, this comparison can also be made for other technologies.

In the calculation of the intrinsic capacitance for each architecture the maximum allowed static error is set at $\frac{LSB}{4}$ (appendix A). The error of the signal on the capacitor is not only determined by the static error, but is also a function of the signal shape and the settling error that is causing a dynamic error. The signal shape is considered to be a continuous analog signal (at the input of the A/D converter or a settling signal which is at maximum a full scale (V_{pp}) voltage step. The bandwidth requirements of these signals are considered to be equal (section 2.3). To guarantee a total error of $\frac{LSB}{4}$, an optimum distribution can be made between the static and the settling error. The required power, when increasing the demands on the static error, is in general a quadratic function of the accuracy increment, while increasing the demands on the settling error is only a weak (logarithmic) function of the accuracy increment. Most of the allowed error of $\frac{LSB}{4}$ is therefore assigned to the static error and the additional power required to reduce the settling error is ignored in the architecture comparison.

It assumed that the converters are able to handle signal frequencies up to the Nyquist frequency. This means that in equation 3.35 f_s of the respective A/D architecture is used. Since not all of the different architectures use a full sample period for signal settling, a normalization factor (x_{NF}) is needed for equal data rate. The flash and folding architectures use a full sample period to drive the calculated intrinsic capacitances. The two-step and pipe-line architectures only have half a sample period available to drive the respective capacitances. This means that the settle time available to charge the intrinsic capacitances is only half of the total sample period and the normalization factor equals two. In the case of the SAR architecture the available settle time is dependent on the number of bits. The SAR architecture requires one cycle for tracking and N cycles for quantization. The normalization factor therefore equals $N + 1$. The normalization factors x_{NF} for each architecture are summarized in table 3.1.

The achieved ENOB at DC from equation 3.35 is assumed to be equal to 2^N for each architecture. As a reference, the capacitance required for sampling a signal at N-bit accuracy is chosen. This intrinsic capacitance is given by equation 3.20 and has a normalization factor of one. For the architecture comparison the following equation is used:

Table 3.1: *The different A/D architectures with their normalizing factors x_{NF}*

Architecture	x_{NF}
Flash	1
Folding	1
Two step	2
Pipe lined	2
SAR	N+1

$$\frac{FoM_{arch\ x}}{FoM_{ref\ SHA}} = \frac{C_{intr,x} \cdot x_{NF}}{C_{intr,ref\ SHA}} \tag{3.36}$$

The result of the comparison is shown in figure 3.32. The reference sample rate is chosen such that the calculated intrinsic capacitance dominates over the parasitic capacitances.

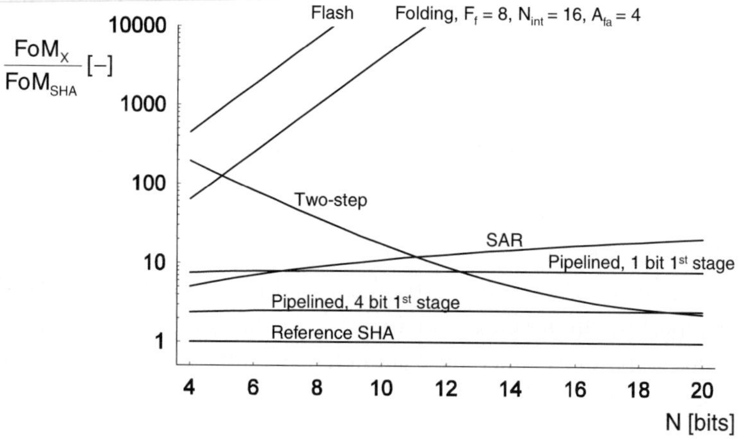

Figure 3.32: *The figure-of-merit of different architectures relative to the intrinsic capacitance of a single sample-and-hold as a function of the number of bits N for 'low' sample rates*

From figure 3.32 it can be seen that without enhancing techniques the flash architecture is the least power-efficient architecture. This is due to the large amount of parallelism of this architecture. For each quantization level an accurate comparator is required. The folding and interpolation architectures reduce the power efficiency by reducing the required number of accurate input transistors by interpolation. Also the number of comparators and their accuracy is reduced, by both folding and folding amplifier gain, respectively. From this figure it is clear that the power of a SHA is negligible in a folding A/D converter. Because of the reduction in both the number of comparators and the accuracy requirement of those comparators by employing residue signal gain, the two-step architecture is able to increase the power efficiency much further. The pipe-line architecture has a *FoM* which remains flat over the whole accuracy range. The intrinsic capacitance of the pipe-line is 8 times larger than the intrinsic capacitance of the single reference sample-and-hold. When the number of bits in the first stage is 4, the total intrinsic capacitance is only 2 times larger. The SAR architecture has a smaller intrinsic capacitor, although the available settling time is much smaller than the settling time of the pipe-line architecture.

From figure 3.32 it can be concluded that making sufficient gain just after the first stage results in the best architectures in terms of power efficiency.

Figure 3.32 shows the *FoM* of the architectures when no enhancement techniques such as interpolation, averaging or offset calibration are used. These enhancement techniques reduce the intrinsic capacitances and therefore increase the power efficiency. For example, when averaging is applied to the flash architecture, reduction of intrinsic capacitance and thus of *FoM* of as much as a factor of 16 can be achieved. The same holds for the two-step architecture, the *FoM* of which can be optimized by reducing the capacitive load of both the coarse and fine converters and the residue amplifier, by averaging, interpolation techniques or calibration. This holds for accuracies below 14 to 16 bits, since above these accuracies the intrinsic capacitance is dominated by the sampling capacitor. The *FoM* of both SAR and pipe-line architectures in figure 3.32 cannot be optimized further by enhancement techniques, since they are determined by the noise requirements.

This comparison is made using the technology parameters of a 0.18 μm technology. Since for high accuracies the intrinsic capacitance of the two-step, pipe-line and SAR architectures are determined by the noise requirements, the difference between these architectures remains the same for different technologies. For low accuracies, the intrinsic capacitances of the flash, folding and two-step architectures are determined by matching requirements. Since for more advanced technologies the required capacitance for matching decreases faster than the required capacitance for noise (section 2.5), the curves of these architectures will decrease with respect to the other architectures.

A/D converter comparison of realizations

The *FoM* of realized A/D converters can also be calculated and compared. However, in the *FoM* calculated by equation 3.35 the input signal amplitude is not considered, while it is obvious from equation 2.6 and equation 2.12 that the input signal amplitude quadratically influences the intrinsic capacitance and thus the power. Optimizing the input signal amplitude to the highest achievable value therefore greatly influences the required power. To be able to compare the real architecture efficiency, the *FoM* of equation 3.35 is normalized to $V_{pp} = 1$ V input signal amplitude, which was also the case in figure 3.32:

$$FoM_{V_{pp}} = \frac{P \cdot V_{pp}^2}{2^{ENOB,DC} MIN(f_s, 2f_{ENOB,DC-0.5})} \tag{3.37}$$

Tables 3.2 through 3.7 show the *FoM* calculated with equation 3.35 and the *FoM* normalized to $V_{pp} = 1$ V using equation 3.37 of realized and published A/D converters from 1998 until 2004. Each table is sorted with the lowest $FoM_{V_{pp}}$ on top. The realized A/D converters are distributed over the categories: flash, folding, two-step/subranging, one-bit-first-stage pipe-line, multi-bit-first-stage pipe-line and SAR. It is difficult to distinguish between the pipe-line and two-step/subranging architecture, when in the realized two-step/subranging architecture re-sampling is also applied (for example [27]). In this comparison the realized converter is put in the two-step category if all of the two (or more) steps consist of more than three bits.

Table 3.2: *Table of realized flash A/D converters*

Ref	N	ENOB	f_s [MS/s]	ERBW [MHz]	P [mW]	*FoM* [pJ]	V_{pp} [V]	$FoM_{V_{pp}}$ [pJV^2]
[41]	6	5,2	22	11	0,48	0,59	0,64	0,24
[42]	7	6,1	450	80	50	4,56	0,8	2,92
[14]	6	5,7	1500	500	328	6,31	1,2	9,09
[43]	6	5,2	400	200	150	10,20	1	10,20
[44]	6	5,8	500	160	225	12,62	1	12,62
[45]	6	5,5	700	136	187	15,19	1	15,19
[13]	6	5,6	900	450	300	6,87	1,5	15,46
[46]	6	5,2	1300	750	545	11,41	1,6	29,20
[47]	6	5,5	75	37,5	110	32,41	1	32,41
[48]	8	7,1	200	50	655	47,75	1,3	80,69

The results of the calculation of the *FoM* normalized to $V_{pp} = 1$ V ($FoM_{V_{pp}}$) from the tables 3.2 through 3.7 and the ENOB dependency of the $FoM_{V_{pp}}$ of the

Table 3.3: *Table of realized folding A/D converters*

Ref	N	ENOB	f_s [MS/s]	ERBW [MHz]	P [mW]	FoM [pJ]	V_{pp} [V]	$FoM_{V_{pp}}$ [pJV^2]
[49]	8	7,5	1600	800	1270	4,38	0,8	2,81
[24]	10	9,2	40	20	65	2,76	1,6	7,07
[50]	8	7,5	600	400	200	1,84	2	7,37
[51]	8	6,7	125	62,5	110	8,46	2	33,86
[52]	8	7,7	10	5	76	36,55	1	36,55
[53]	7	6,4	300	70	200	16,92	1,6	43,31

Table 3.4: *Table of realized two-step/subranging A/D converters*

Ref	N	ENOB	f_s [MS/s]	ERBW [MHz]	P [mW]	FoM [pJ]	V_{pp} [V]	$FoM_{V_{pp}}$ [pJV^2]
[27]	14	12	100	50	1250	3,05	2	12,21
[54]	8	7,6	100	50	118	6,08	1,6	15,57
[55]	10	9,7	20	10	75	4,51	2	18,03
[56]	12	10,3	50	25	850	13,48	1,6	34,52
[28]	**10**	**9,1**	**25**	**12,5**	**195**	**14,21**	**1,6**	**36,39**
[57]	13	11,1	40	15	800	12,15	2	48,60
Chap 7	**16**	**10.8**	**20**	**2**	**141**	**19.7**	**2**	**79.1**
[58]	**12**	**10,3**	**54**	**4**	**295**	**29,25**	**1,8**	**94,77**

reference sample-and-hold capacitance are plotted in figure 3.33, with the realized ENOB on the x-axis.

From figure 3.33 it can be seen that the flash architecture is (barely or) not used at all for accuracies above 8 bits. This architecture is not suitable for high accuracies because of the large intrinsic capacitance required. However, for low accuracies it can be favorable to use interpolation and offset compensation techniques to reduce the intrinsic capacitance which optimizes the efficiency, which is shown by [41].

The realizations of the folding architecture show better efficiency than the realized flash architectures, as expected from the intrinsic load capacitances. The realization described in [24] shows a good efficiency. This is the result of the use of a BiCMOS technology, which has better matching properties, resulting in a lower intrinsic capacitance.

The realized $FoM_{V_{pp}}$ of the two-step or subranging architectures is relatively constant for different values of the ENOB. In figure 3.32 it was shown that the $FoM_{V_{pp}}$ of a two-step converter increases less rapidly than the noise-limited architectures. Furthermore, because the two-step architecture is matching limited, calibration

Table 3.5: *Table of realized pipe-line A/D converters*

Ref	N	ENOB	f_s [MS/s]	ERBW [MHz]	P [mW]	FoM [pJ]	V_{pp} [V]	$FoM_{V_{pp}}$ [pJV^2]
[59]	10	9,2	220	100	135	1,15	1	1,15
[60]	8	6,9	4000	800	4600	24,07	0,25	1,50
[61]	8	6,5	20000	2000	9000	24,86	0,25	1,55
[62]	10	8,7	140	70	123	2,11	1	2,11
[63]	10	8,7	20	10	20	2,40	1	2,40
[64]	10	9	30	15	16	1,04	1,6	2,67
[65]	10	9,2	100	50	180	3,06	1	3,06
[66]	10	8,7	30	15	16	1,28	1,6	3,28
[34]	10	9,4	14,3	4	36	6,66	0,8	4,26
[67]	10	9,4	80	40	69	1,28	2	5,11
[68]	8	7,6	500	250	950	9,79	0,75	5,51
[69]	15	12	20	8	233	3,56	1,5	6,97
[70]	12	11,1	20	10	250	5,69	2	22,78
[71]	12	10,8	10	5	338	18,96	1,6	48,53
[72]	12	10,9	61	2,5	600	62,80	1,4	123,09
[73]	14	12	20	10	720	8,79	4	140,63
[74]	14	12,8	5	1	320	22,44	4	358,97

Table 3.6: *Table of realized pipe-line A/D converters with a multi-bit first stage*

Ref	N	ENOB	f_s [MS/s]	ERBW [MHz]	P [mW]	FoM [pJ]	V_{pp} [V]	$FoM_{V_{pp}}$ [pJV^2]
[33]	14	12	75	37,5	341	1,11	2	4,44
[36]	14	11,9	75	37,5	340	1,19	2	4,74
[75]		12	50	23	780	4,14	1,2	5,96
[76]	8	7,5	150	80	71	2,61	1,6	6,69
[37]	12	11	75	37,5	290	1,89	2	7,55
[30]	10	9,5	40	20	70	2,42	2	9,67
[35]	12	11,3	65	32,5	480	2,93	2	11,72
[77]	14	11,9	10	15	112	2,93	2	11,72
[78]	15	11,7	40	19	394	3,12	2,25	15,78
[79]	12	11	105	52,5	850	3,95	2	15,81

can be applied to reduce the intrinsic capacitance. This is discussed in the next chapter.

As can be seen in figure 3.33, the $FoM_{V_{pp}}$ of the pipe-line converter increases as a function of the realized ENOB. The slope of the line through the realizations with

Table 3.7: *Table of realized SAR A/D converters*

Ref	N	ENOB	f_s [MS/s]	ERBW [MHz]	P [mW]	FoM [pJ]	V_{pp} [V]	$FoM_{V_{pp}}$ [pJV^2]
[80]	9	8,2	0,15	0,03	0,03	1,70	0,5	0,43
[81]	8	7,7	0,1	0,045	0,0046	0,25	1,4	0,48
[82]	8	7,2	20	10	12,9	4,39	1	4,39
[83]	13,5	13	0,016	0,001	0,15	9,16	2,4	52,73
[84]	12	11,5	0,125	0,0625	16	44,19	4	707,11

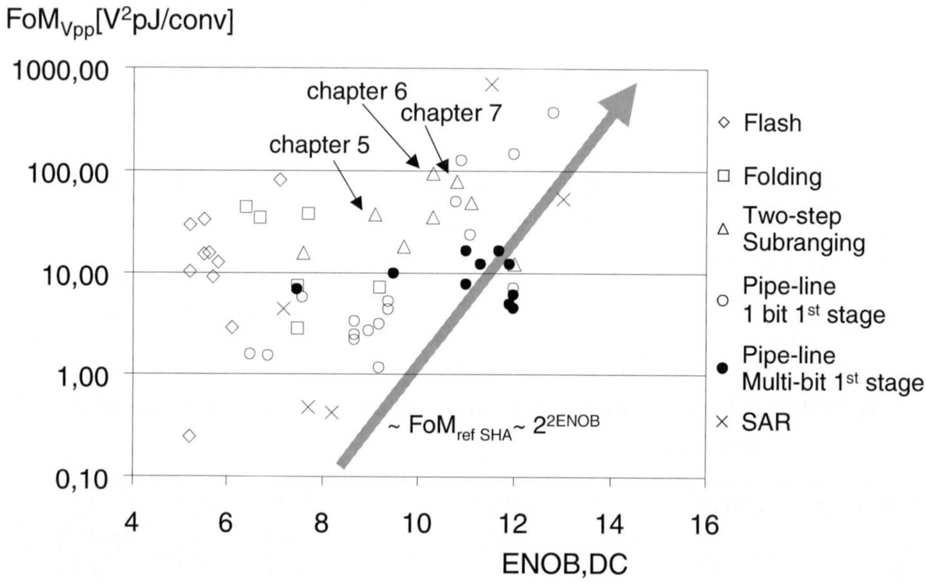

Figure 3.33: *The figure of merit normalized to $V_{pp} = 1$ V of realized A/D converters and the slope of a reference SHA as a function of the achieved ENOB*

low $FoM_{V_{pp}}$ is equal to the slope of the $FoM_{V_{pp}}$ of the reference sample-and-hold capacitance. This is also equal to the expected slope shown in figure 3.24. The converters on this line are designed with capacitors, limited by the noise requirements. Calibration does not reduce the capacitance in noise-limited designs. The use of a multi-bit first stage does, however, reduce the intrinsic capacitance with respect to the one-bit first stage pipe-line converter, as is shown in figure 3.33.

An exception to this is the realization presented in [69], where a pipe-line A/D converter with a one-and-a-half-bit first stage achieves almost the same efficiency as the pipe-line A/D converters with a multi-bit first stage. This is a remarkable result.

The SAR architecture is also a noise-limited design. This is shown in figure 3.33 by the increasing $FoM_{V_{pp}}$ as a function of the ENOB.

The result of the comparison of the realized A/D converters is not exactly equal to the theoretical result shown in figure 3.32. This is caused by the fact that the result from figure 3.32 shows the bare minimum achievable $FoM_{V_{pp}}$, while the realized $FoM_{V_{pp}}$ depends very much on the actual implementation. However, figure 3.33 confirms the theory derived from figure 3.32 that for higher accuracies, the realizations of the two-step/subranging or pipe-line with multi-bit first stage A/D architectures are most efficient with respect to the other architectures. For high accuracy the number of levels and the associated gain of the first stage have to be large for high efficiency. In this case the dominant error source with respect to noise and matching is the first stage. After this first stage the contributions of the errors to the overall error are greatly reduced.

> An important observation from figure 3.32 is that the *FoM* is not independent of the accuracy for a given architecture, but increases for higher accuracy. This is caused by the fact that the capacitive load of the capacitance determined by both noise and matching increases with 2^{2N}, while the achieved accuracy only increases with 2^N.

3.7.3 Architecture comparison as a function of the sampling speed

Figure 3.32 shows the differences between the architectures for low sample rates. When the power is increased, indicated by $mult$ in figure 2.5, the available settle time decreases by the same amount, allowing an increase of the sample rate. When the power and sample rate are increased simultaneously, the *FoM* remains constant. However, when $mult$ reaches the value where the available settle time does not decrease linearly with $mult$, the power increases more rapidly than the sample rate. This means that the *FoM* increases and the converter becomes less power efficient.

This available settling time is used differently for the several A/D converter architectures. As can be seen in table 3.1, the flash and folding architectures require only one settle period for one conversion cycle. The maximum achievable sample rate is therefore equal to $\frac{1}{t_{s,min}}$, where $\frac{1}{t_{s,min}}$ is determined by the minimum achievable time constant τ_{min} (section 2.4). The two-step/subranging and pipe-line architectures use two settle periods within one conversion cycle, which results in a maximum sample rate of $\frac{1}{2t_{s,min}}$. The SAR architecture requires $N+1$ settle periods for one conversion cycle, resulting in a maximum sample rate of $\frac{1}{(N+1)t_{s,min}}$. Non-idealities, such as necessary switching time in the two-step, pipe-line and

SAR architectures or non-overlapping clock generation, which require part of the total sampling period, are not taken into account. This effect reduces the maximum achievable sample rate of these architectures even further.

3.8 Conclusions

In this chapter the most important A/D converter architectures for high speed and high bandwidth are compared with respect to power consumption and maximum achievable speed. From the combination of figure 3.32 and the maximum achievable sample rates it can be concluded that for low sample rates the architectures which reduce the number of necessary accurate devices are much more power efficient than the parallel type of converters like the flash and the folding architectures. This is especially the case for high accuracies, where the size of the intrinsic capacitance of the flash and folding architecture becomes impractical. However, for high sample rates parallel architectures are more power efficient since the conversion is carried out in only one step.

Practical realizations also show the trend that for a high sample rate the flash and folding are the optimum architectures. However, due to the parallelism, the accuracy is limited to 8/10 bits. For the higher accuracies two-step/subranging and pipe-line with multi-bit first stage architectures are more efficient.

Designing for high accuracy requires some kind of calibration. When more bits are applied in the first stage, the accuracy of the generation of the least significant bits becomes less critical. Generating more bits in the first stage(s) thus reduces the calibration to only this stage. This is in contrast to the more parallel type of architecture where all the decision blocks need calibration separately.

Because of its power efficiency for high resolution, the two-step architecture is further analyzed in the remainder of this book. The next chapter discusses enhancement techniques for this architecture to increase the accuracy without increasing the power consumption, thereby increasing the power efficiency.

Chapter 4

Enhancement techniques for two-step A/D converters

4.1 Introduction

In the previous chapter the two-step architecture was chosen for further investigation. This chapter first discusses the error sources in a two-step architecture. The major accuracy challenge in the two-step architecture is the need for an accurate gain factor between the quantization stages to fit the residue signal exactly in the fine sub-range. In [3] an architectural solution for this requirement is provided, which is explained in section 4.3. Due to this solution, the offset of the residue amplifier stage becomes very important. Section 4.4 gives an overview of the existing offset calibration methods as they are applicable on different A/D converter architectures. In section 4.5, a solution for background on-line digital offset extraction and analog compensation is discussed.

4.2 Error sources in a two-step architecture

Figure 4.1 shows the general diagram of the two-step architecture (section 3.4). Added to this figure are the main error sources which can limit the performance of the two-step architecture.

Each of the error sources shown in figure 4.1 will be discussed and solutions are provided in order to limit the performance reduction. The sample-and-hold is not considered since the requirements of this block in the two-step architecture are not different from the requirements in other architectures.

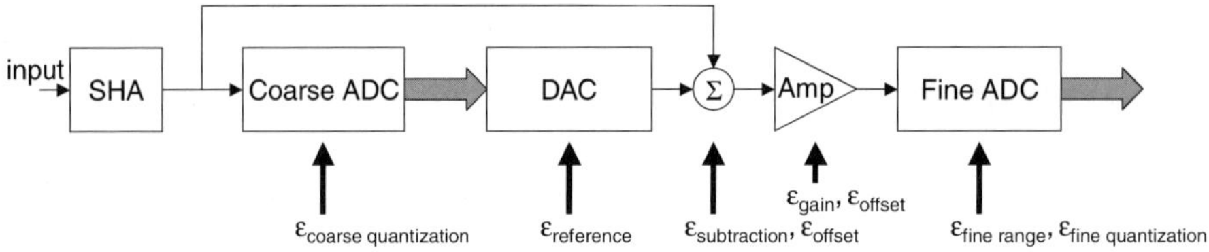

Figure 4.1: *Diagram of the general two-step architecture with the possible error sources*

- The coarse quantization determines the selection of the subranges and the DAC setting. When an error is made in this quantization ($\varepsilon_{coarse\ quantization}$), the wrong subrange is selected, which results in missing codes, as shown in figure 3.18. By adding over-range to the fine A/D converter (section 3.4) the accuracy requirements of the coarse A/D converter are reduced significantly.

- The references of the D/A converter are important since they determine the accuracy of the total A/D converter. An error in the references ($\varepsilon_{reference}$) results in an INL error. The reference therefore determines the accuracy of the overall A/D converter. In general, the matching of resistances provided in CMOS technology is sufficient [85, 86] for more than 12-bit accuracy and is adequate for the scope of this book.

- The subtraction of the input signal and the D/A converter output determines the overall achievable accuracy. An offset or a gain error in the subtraction ($\varepsilon_{subtraction}$) results, for example, in an INL error. The consequences of these errors are similar to those of the errors made in the residue amplifier, which will be described later.

- The residue amplifier is an important accuracy-determining element of the two-step architecture. An offset (ε_{offset}) on the residue amplifier or on the subtractor gives a DC shift of the fine A/D converter reference with respect to the coarse A/D and D/A converter range. When over-range is applied in the fine A/D converter (section 3.4), this offset can be larger than without over-range. However, an error in the gain factor of the residue amplifier (ε_{gain}) is much more critical [87]. A gain error in the cascade of the subtractor and the residue amplifier causes the total range of residue signal, which is the result of the subtraction of the input signal and the D/A converter signal, not to fit in the fine A/D converter range. This may lead to non-monotonicity or missing codes [87]. The next section (4.3) discusses an architectural solution to overcome this limitation.

- An error in the range of the fine A/D converter ($\varepsilon_{fine\ range}$) results in an error similar to a residue amplifier gain error. The gain of the subtractor and amplifier should therefore be linked with the fine A/D converter range. This is explained in the next section. The accuracy of the fine quantization determines the overall accuracy of the A/D converter. However, since the residue amplifier provides gain, the accuracy requirements are reduced by this gain factor.

- The fine quantization determines the achievable accuracy of the total A/D converter. An error in this quantization ($\varepsilon_{fine\ quantization}$) increases the DNL of the total A/D converter.

Most of the error sources described above are dealt with in section 3.4. The most critical point in the two-step architecture is the interstage gain factor, which will be discussed in the next section and improvements proposed.

4.3 Residue gain in two-step A/D converters

4.3.1 Single-residue signal processing

The two-step A/D architecture from figure 3.16 generally works with single residue signals. An implementation of the single residue two-step converter is shown in figure 4.2.

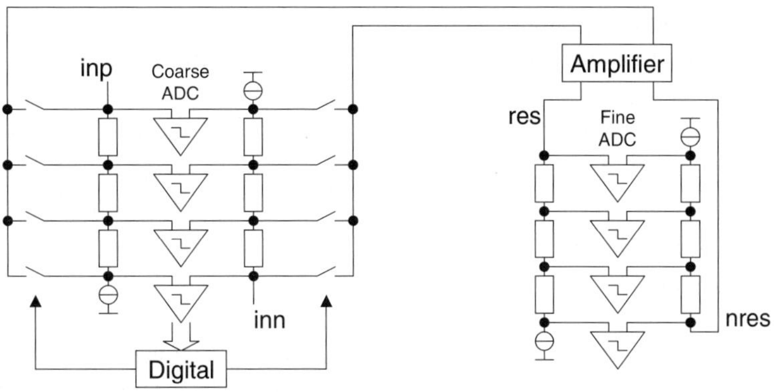

Figure 4.2: *Two step architecture with single residue system*

The coarse converter part is similar to the flash architecture described in section 3.2. According to the coarse decision, switches are set to tap the proper residue signals from the moving resistor ladders. This is illustrated in figure 4.3.

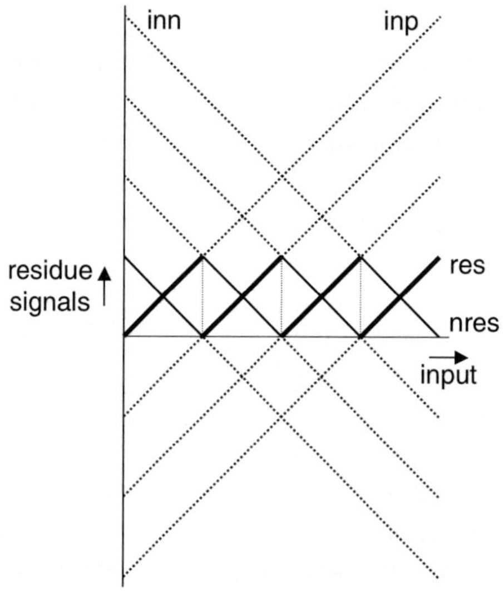

Figure 4.3: *Residue signal generation in a single residue system (with a residue amplifier gain of 1)*

For illustration purposes the residue amplifier gain in figure 4.3 is set at 1. In practice the gain is higher than 1. These residue signals are applied to a differential amplifier, which applies its output to two resistor ladders which generate the fine bits similar to the coarse converter. The amplitude of the residue signals, *res* and *nres*, applied to the fine ladders is determined by the resistance of a coarse tap, the current in the coarse ladder and the gain of the residue amplifier.

The range of the fine A/D converter is determined by the current in and the resistance of the resistor ladders of the fine converter. Due to deviations in one of the elements mentioned above, ranging errors in the fine converter are easily made. Figure 4.4b shows the errors which occur when the gain of the residue amplifiers is less than expected.

The bold lines, *res* and *nres*, are the signals which are applied to the tops of both resistor ladders. Because the gain of the residue signals is too low, the signal is too small for the fine range. This can be seen at the beginning and at the end of a subrange in figure 4.4b. This misfit causes jumps in the transfer curve of the total A/D converter at the subrange transitions, which leads to missing codes or non-monotonicity. To circumvent this source of errors several precautions can be taken. For example, if the fine range matches the full-scale residue amplifier output signal swing, which means that the references for fine converter experience

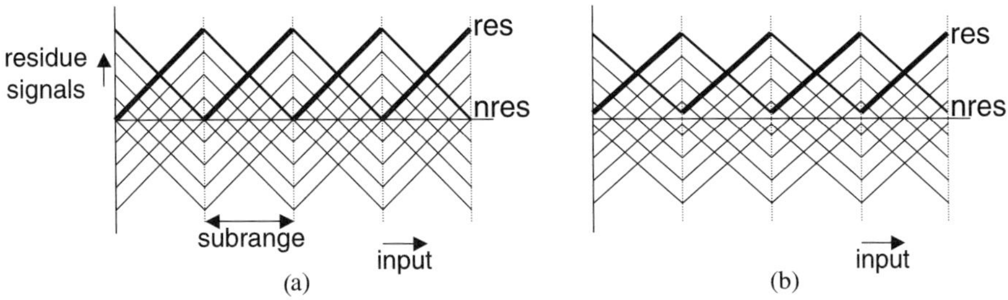

Figure 4.4: *Fine converter signals of a single residue system, with nominal amplifier gain (a) and with less than nominal gain (b)*

the same gain as the residue signal, the problem described above can be reduced [88, 89, 90]. However, this still limits the achievable accuracy to about 10 bits.

4.3.2 Dual-residue signal processing

Since only one side of the fine resistor ladders are determined by the residue signals and the fine range is determined by a separate resistor value and current in the single residue configuration, monotonicity cannot be guaranteed by design. The other side of the fine resistor ladder can also be connected to a separate residue amplifier, which is connected to one tap shifted on the coarse resistor ladder. In this case, the fine range exactly fits the residue signal [3]. This is illustrated in figure 4.5.

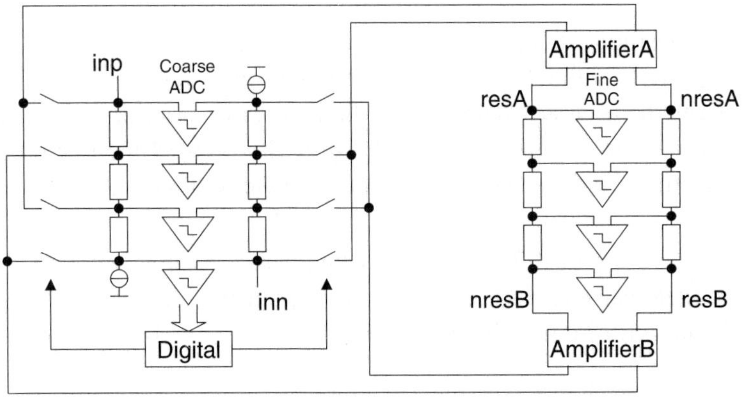

Figure 4.5: *Two-step architecture with dual-residue system*

The dual-residue system from figure 4.5 can also be considered as two amplifiers generating zero-crossings at the edges of the subrange and, by resistive interpolation, the additional required zero-crossings for the comparators are generated. The switches are connected in such a way that, when switching from one subrange into the next subrange, one amplifier remains connected to the coarse ladders, while the other amplifier changes two tap positions [91]. With this improved switching, even with amplifier errors as described later, continuity and monotonicity are guaranteed. As a result of not only driving the top side, *resA* and *nresA*, of the fine reference ladders, but also the bottom side, *resB* and *nresB*, from the coarse resistor ladder, the range of the fine A/D converter is made directly dependent on the selected coarse subrange and the amplifier gain. If in this case the gain of both residue amplifiers is reduced, it does not result in an error of the fine conversion. This is due to the fact that the fine range is dependent on the gain of the residue amplifiers. This is illustrated in figure 4.6.

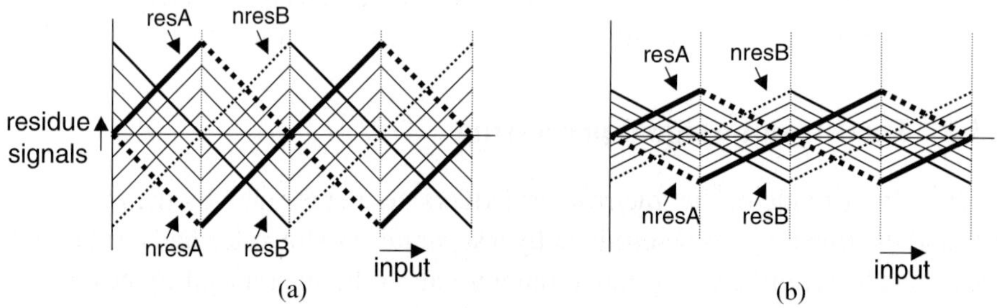

Figure 4.6: *Residue signals of a dual-residue system, with nominal amplifier gain (a) and with less than nominal gain (b)*

As can be seen in figure 4.6, the zero-crossings of the fine comparators remain at the same position, independently of the gain. The absolute gain of the two residue amplifiers is therefore not important. However, a lower residue amplifier gain increases the accuracy demands on the fine conversion. A difference in gain between the residue amplifiers causes the zero-crossings of the fine comparators to deviate from the ideal value. This effect is shown in figure 4.7.

As can be seen in figure 4.7a, the interpolated zero-crossings are shifted with respect to the ideal value and this causes an error in the transfer of the A/D converter. The maximum error in the INL curve from figure 4.7b can be calculated by:

$$INL_{max,fine} = \frac{\sqrt{G_R} - 1}{\sqrt{G_R} + 1} \cdot 2^{N_F}, \tag{4.1}$$

with $G_R = \frac{A_A}{A_B}$, the gain ratio of the gain of amplifier A and amplifier B and N_F is the number of bits of the fine A/D converter. By rewriting equation 4.1 the maximum relative gain error $|G_{R,max} - 1|$ can be calculated. Assuming

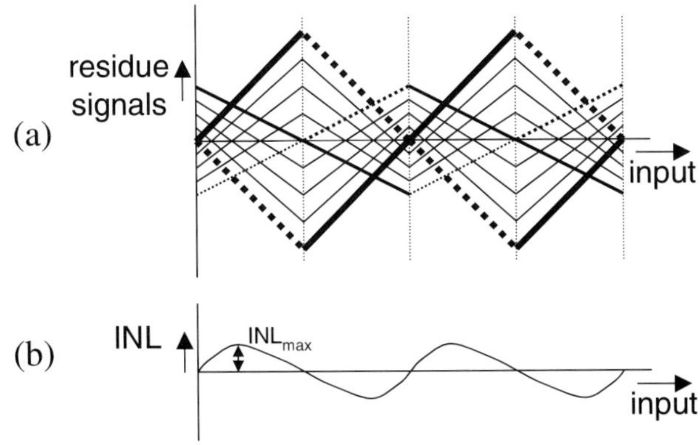

Figure 4.7: *Residue signals of a dual-residue system, with different gain for both amplifiers (a) and the respective INL (b)*

that the maximum allowed INL is $\frac{LSB}{4} = \frac{1}{2^{N_F+2}}$ for less than 0.4 loss in ENOB (figure A.1), then the maximum relative gain error is given by:

$$G_{R,max} = \frac{\left(2^{N_F+2}+1\right)^2}{\left(2^{N_F+2}-1\right)^2} \tag{4.2}$$

In figure 4.8 the maximum relative gain error as a function of the required number of bits N_F in the fine A/D converter is shown.

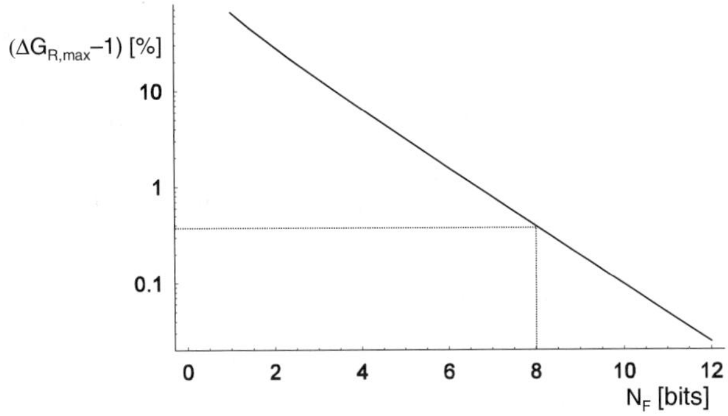

Figure 4.8: *Maximum relative gain error as a function of the accuracy*

The gain matching requirement is exponentially dependent on the number of bits in the fine converter. When, for example, the fine A/D converter has to resolve 8 bits ($N_F = 8$), the required gain matching of the two residue amplifiers has to be better than 0.4%. This matching is achieved by the intrinsic matching of the technology. The gain values of the amplifiers have to be calibrated if more matching is required.

Another point for attention is the offset between the two residue amplifiers. Due to mismatch, the zero-crossings of the residue amplifiers can be shifted with respect to the wanted position. A difference between the offsets of the residue amplifiers results in an INL error. This is illustrated in figure 4.9a.

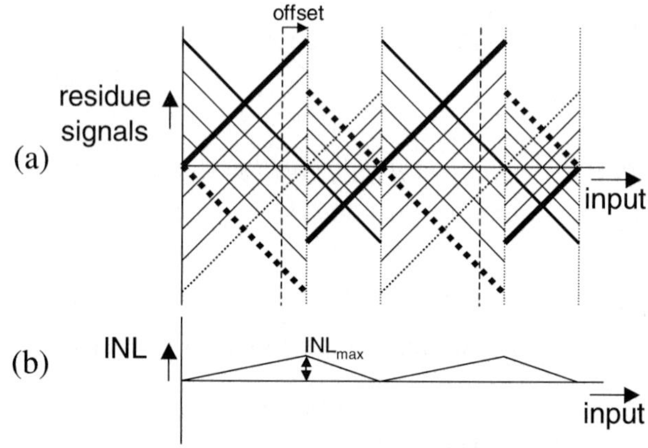

Figure 4.9: *Effect of residue amplifier offset on fine A/D converter zero-crossings (a) with the respective INL (b)*

As can be seen in figure 4.9b, the INL curve follows a continuous line and there will be no large code transitions due to the dual residue signal processing in the case of amplifier offset. This is a result of interpolation between the two residue amplifier outputs, which determine the outer ends of the range. This means that the offset is spread over the whole range, the maximum DNL is therefore the difference between the input-referred offsets of both amplifiers divided by the resolution of the fine A/D converter (equation 4.3). The maximum INL, however, is equal to this offset difference (equation 4.4).

$$DNL_{max} = \left| \frac{V_{offset\ A} - V_{offset\ B}}{2^{N_F}} \right| \tag{4.3}$$

$$INL_{max} = \left| V_{offset\ A} - V_{offset\ B} \right| \tag{4.4}$$

To achieve an INL better than ±0.25 LSB, the difference in the amplifier offset has to be within ±0.25 LSB.

4.3.3 Conclusions

Dual-residue signal processing [3] is a powerful technique to circumvent ranging errors which occur when the coarse subranges, residue amplifier gain and fine range are not coupled to each other. When improved switching is used, continuity and monotonicity are guaranteed. Dual-residue signal processing spreads the errors of the amplifiers over the whole fine range, which results in an improved DNL, and no missing codes will occur. A gain error is also spread over more codes instead of causing missing codes. However, the difference in offset of the two residue amplifiers determines the INL of the overall A/D converter. For high accuracy this means that the input devices of the amplifier need to be large to ensure a sufficiently low offset. The next sections deal with offset calibration techniques, which allow the use of small input devices to improve the power and speed of the converter.

4.4 Offset calibration

4.4.1 Introduction

In the single residue architecture, the offset of the residue amplifier is not critical. The over-range of the fine A/D converter is able to compensate the offset error. When dual-residue signal processing is used in a two-step architecture the offset of the two residue amplifiers determines the INL, as described in the previous section. The offset of the residue amplifier stage needs to be calibrated in order to prevent the use of large input devices[1]. This section describes a number of calibration techniques which can be applied to the two-step architecture.

4.4.2 Calibration overview

The offset of residue amplifiers in two-step converters, or in general in the critical blocks of A/D architectures, can be calibrated in order to reduce the intrinsic capacitance, thus reducing the required power. These critical blocks are, for example,

[1] Another option could be to calibrate the gain of a single residue system instead of the offset in a dual-residue system. However, in this book offset calibration is chosen over gain calibration since offset can be measured and compensated more easily because it is a DC effect.

the comparators or the amplifier stages. The calibration can be performed at several 'moments' during the lifetime of an A/D converter, as shown in figure 4.10.

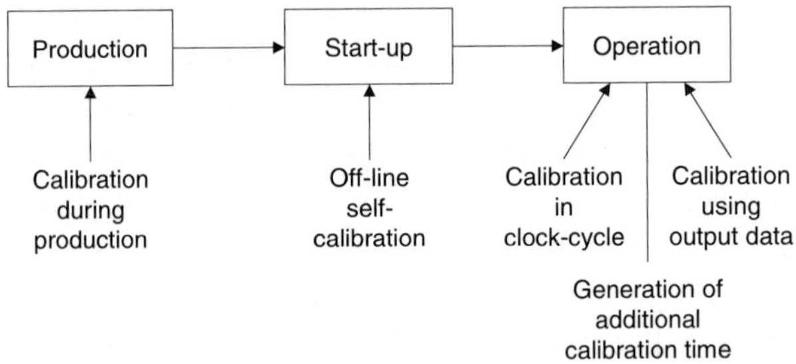

Figure 4.10: *Calibration moments during the lifetime of A/D converters*

These 'moments' occur during fabrication (testing), at start-up or continuously during operation. Depending on when this 'moment' occurs, additional hardware or calibration time, etc., is needed. These calibration schemes can be carried out either in the analog or in the digital domain. The following paragraphs describe the different possibilities and their advantages and limitations.

Calibration during production

The most straightforward method of calibration is carried out off-line during fabrication. When this method of calibration is used the process needs a type of programming facility on-chip, like laser-trimming or -cutting, or read-only memory (ROM) programming. These values are used for compensating the errors in the analog or in the digital domain.

- A current D/A converter which is used for generating the references in, for example, a two-step A/D converter can be calibrated by adjusting each of the separate current sources to their wanted value. This can be done by using additional compensation D/A converters for generating the compensation current [92, 27]. These compensation D/A converters are programmed during fabrication, for example by laser cutting or trimming. This can also be used to generate a compensation current for the offset current which is generated in a differential pair due to mismatch [93]. Also other analog properties like capacitor values can be calibrated by using the programming possibility during fabrication [79]. The advantage of laser trimming a circuit is that after trimming the errors of the analog circuitry are removed, resulting in a ‘perfect’ circuit.

- The correction of the errors generated in the analog domain can also be performed in the digital domain [35, 94, 95]. In this case during fabrication (or testing) the errors of the A/D converter are measured in the digital domain. The correction numbers are calculated and the correction data is programmed during a fabrication step in a programmable read-only memory (PROM). When the A/D converter is used, the analog part generates raw data, which contains errors caused, for example, by capacitor mismatch or transistor mismatch. This raw data is used to estimate the corresponding error. This error is then subtracted from the raw data, which results in the corrected output data. As in the analog trimming case, when the raw data is sufficiently accurate not to influence the quantization error, the resulting A/D converter is ‘perfect’, without static errors.

- To overcome calibration errors when the offsets change due to stress after packaging, both types of calibration during fabrication can be carried out by electrically-programmable read-only memory (EPROM) programming during the final test (after packaging) [96].

The disadvantage of calibration during fabrication is that an additional processing or fabrication step or testing time is required. Each individual A/D converter needs to be characterized separately, and the individual errors need to be corrected. This involves a lot of additional and expensive testing time. And since the calibration is performed only once, errors which are caused, for example, by aging or temperature-dependent behavior are not removed.

Off-line self-calibration

The analog error correction can be stored in a (P)ROM at fabrication, but when a random-access memory (RAM) is used an additional IC-processing step is not required. The RAM is programmed during a calibration cycle at start-up, but this can also be done during operation in a time interval when the A/D converter is not used in the system. For example, during the line blanking in a video system. This type of calibration is generally known as self-calibration (or calibration at start-up) [97].

- During self-calibration the A/D converter is disabled, which means that the analog input signal is not converted. Instead, a known input signal is applied or the A/D converter is put in a known state, such that the output of the A/D converter is a known digital signal. Due to mismatches in the components used, e.g. capacitors [97, 98, 99, 100, 101, 70, 102], the resulting digital signal deviates from the wanted signal. The difference in the wanted and the resulting digital signal is used to calibrate the errors. The calibration can be performed by shifting the references that are applied during conversion [97], which are dependent on the raw data generated by the A/D converter. Another manner of calibrating the errors in the analog domain in, for example, a switched capacitor implementation is by trimming the capacitors [103]. However, the calibration can also be performed completely in the digital domain [98, 99, 100, 101, 70, 102]. The code-dependent correction terms, which are also derived by applying known signals to the A/D converter, are stored in a digital register or a ROM. During normal operation these correction terms are added to the conversion results to calibrate the errors.
- A more flash-type converter-specific calibration method involves assigning more than one comparator to each level [43]. The trip-points of the comparators are distributed because of transistor mismatch. By selecting the comparators closest to each trip-point, and powering down and ignoring the others, the DNL can be reduced significantly. However, this method relies on there being at least one comparator with the proper trip-point available for each decision level. This can only be satisfied when sufficient comparators are available for each decision point, which requires an overhead in silicon area.

Calibration at start-up offers the advantage over calibration during fabrication because each time the A/D converter is powered up, or even during a non-operation time period, the A/D converter is calibrated. This means that mismatch effects due to aging are compensated. But the calibration has to be performed repetitively ‘upon request’ to ensure that mismatch effects like temperature drift or supply variations are compensated for. This requires a system-dependent calibration scheme. Calibration in the analog domain results in an A/D converter with nearly perfect analog circuitry. However, calibration in the digital domain requires sufficient resolution to ensure calibration to a sufficiently low INL of the A/D converter [98]. This is necessary because, due to digital adding and subtraction, quantization errors are generated. Using more bits (typically one or two) in the calibration process and truncating the raw data to N bit reduces this effect sufficiently. This does, however, mean that for an N-bit A/D converter, $N+1$ or $N+2$ bit raw data has to be generated by the A/D converter, which involves additional power and area.

Calibration in clock cycle

Instead of calibrating the A/D converter only during fabrication, at start-up or at a calibration request, the calibration can take place every clock cycle.

- During sampling of the signal the offset is measured and stored on a capacitor, while during the hold (or amplifying) period this offset value is subtracted from the input signal, which results in cancellation of the offset [26, 104, 18, 55].
- Correlated double sampling [105] cancels the offset by first sampling the input signal plus the offset. During the next phase only the offset is sampled on an additional capacitor. The third phase amplifies the input signal, thereby canceling the offset. This can also be used in a comparator stage [1] by first sampling the input signal on sampling capacitors. The second phase generates a signal which is proportional to the input signal plus offset and a signal which is proportional to only the offset. These values are subtracted and in the last phase this resulting signal is latched, which generates the comparator output signal.
- By using an auxiliary amplifier, the offset of an amplifier can be compensated without using capacitors in the signal path [106, 28]. During the compensation phase the inputs of the main amplifier are short circuited, while the offset calibration loop is closed and the offset is sampled on the compensation capacitors. When the main amplifier is used in amplify mode, the auxiliary amplifier with the main amplifiers offset at the input cancels the offset of the main amplifier. The compensation phase is performed every clock cycle.

The main advantage of these techniques is that the offset is calibrated every clock cycle and changes over time are therefore calibrated as well. However, the major disadvantage is that the available settle time for the signal processing phase is halved, since the other half is used for the offset canceling phase. In the case of correlated double sampling the settle time is even three times smaller. In general this means that the power is increased by the same factor.

Generating additional calibration time

The previous paragraph shows that calibration can be carried out on-line, but that this involves additional power. There are several calibration methods that can be performed on-line to reduce the settling requirement back to a full sample period without this involving additional settling speed.

- The most straight-forward method is to replace the part of the converter which has to be calibrated [71] with a circuit which takes over the functionality. This can be executed to the extreme if a complete A/D converter is replaced by another A/D converter when it has to be calibrated [107]. This is advantageous when more than one A/D converter is used time-interleaved in order to increase the overall sample rate. These additional stages require some additional power, although, this overhead becomes less as the number of stages (or complete A/D converter channels) which have to be calibrated increases.

- The circuit to be calibrated can also be taken from the system without replacing it. The missing signals can be generated (by interpolation, for example) from the remaining circuitry [108, 57]. In this way no missing samples in time occur and no additional circuitry is required. But to ensure a sufficient quality of interpolation, the linear region of the remaining circuitry needs to be large enough, requiring additional power. This analog interpolation and calibration is due to the required parallelism only applicable for flash-type converters.

- It is possible to generate some additional calibration time without the use of extra circuitry or sampling at a higher sample rate by using a skip-and-fill algorithm [109]. The calibration is based on the calibration described in the off-line self calibration section; this is not performed at start-up or on request, but by skipping a normal conversion cycle and using this time for calibration. The skipped sample is filled by non-linear interpolation of the digital output data of the converter. The skipped conversion cycle is chosen at random. However, this implies that no signal frequencies above $\frac{1}{3}f_s$ can be applied to the converter in [109] because then the interpolation algorithm does not function properly. Other interpolation algorithms can be chosen to circumvent this restriction.

- The additional calibration time can also be generated by using queue-based sampling [84]. The input signal is sampled at uniformly distributed time intervals, and the output signals are re-sampled with another sample and hold, at a higher sample rate. The output of this sample and hold is quantized with an A/D converter running at this same higher sample rate. When the queue formed by the second sample and hold is empty, the A/D converter has extra time for a calibration cycle. The A/D converter runs at a slightly higher sample rate and therefore requires additional power.

The calibration algorithms described above all work on-line, although to some extent additional overhead in speed, circuitry or linear region is required and this does involve some additional power. In the case of the skip and fill algorithm the complete Nyquist band cannot be handled by the A/D converter.

Calibration by using output data

To reduce the overhead of additional circuitry even further than described in the previous section, the offset information can be extracted from the digital output data of the converter. Since the output data contains both signal and offset errors (or other errors such as gain mismatch), the offset needs to be distinguished from the input signal.

- Data-dependent shuffling of capacitors relaxes the requirement of matching of capacitors in terms of missing codes or non-monotonicity [110]. Even in the case of capacitor mismatch, the DNL is improved significantly, although the INL is still determined by the mismatch. By choosing a pseudo random-dependent shuffling instead of data-dependent shuffling, even the INL errors are averaged out [111]. They still exist on sample to sample base, which does not improve the SNR, although over more samples the errors are averaged to zero. Since the signal which shuffles the capacitors is generated in the digital domain, this information can be used to estimate the error of the capacitors in the digital domain. The digital estimate of the capacitance error can be used to compensate digitally for the error. This is, however, not yet proven as a complete system [111].
- By adding a pseudo random signal to the analog input [112, 113], which is sufficiently un-correlated with the input signal, the gain errors between two (or more) interleaved A/D converters can be extracted. The added signal is converted by the A/D converters together with the input signal. In

the digital domain this added signal is subtracted. When there is gain mismatch, a residual pseudo random signal remains in the output of the A/D converters. This residual signal is correlated with the pseudo random signal and accumulated in an integrator. Due to the integration in the accumulator and correlation with the pseudo random signal, the input signal is averaged out. The resulting value in the accumulators is used to adjust the gain of the A/D converters by multiplying in the digital domain. The disadvantage of this method is the required additional input signal range for adding the pseudo random signal. This can be as large as one quarter of the full range. In this case detection of errors to an N-bit level requires an $N+2$-bit A/D converter [112], in order to detect properly these added pseudo random signals. The offset between the two A/D converters in [112] is removed by accumulating the difference of the outputs of both A/D converters. Due to offset, the content of this integrator differs from zero, which is a measure for the offset. However, a signal at $\frac{f_s}{4}$ is also seen as offset, which disturbs the offset cancellation.

In general, for all the calibration methods, the maximum achievable accuracy is dependent on the accuracy of the error-compensating mechanism. In the digital domain this is determined by the number of bits used for the compensation before the data is truncated to the number of output bits of the A/D converter. In the analog domain, this can be determined, for example, by the resolution of the D/A converter that provides the compensation value. When capacitors are calibrated, the value of the smallest capacitor determines the accuracy.

4.4.3 Conclusions

All of the calibration techniques discussed, except those in the last section, require calibration time. This is subtracted from the total conversion time of the A/D converter. It is obvious that calibration at start-up requires additional time when the A/D converter is switched on. After that, the A/D is calibrated and acts as an ideal converter but, as with calibration during fabrication, changes in time due to temperature or supply changes are not calibrated. The offset measuring and calibrating during each sample takes additional time during conversion. The calibration algorithms which work with the output data of the converter have not yet been proven to work in order to improve the performance of a single A/D converter or require an additional input signal range for the error detection. The next section describes an algorithm which uses the digital data to extract the errors and continuously calibrates the A/D converter. The errors are calibrated in the analog domain, resulting in an A/D converter with almost perfect analog circuitry. This algorithm uses only very few additional non-critical analog-components.

4.5 Mixed-signal chopping and calibration

This section explains a calibration algorithm which extracts residue amplifier offsets from the normal processed data by the converter without taking additional time. The amplifier offset which has been identified is corrected for the analog domain, which means that the resulting calibrated A/D converter has residue amplifiers without offsets.

4.5.1 Residue amplifier offset chopping

Figure 4.11 shows the residue amplifier with quantizer at the output. If this amplifier comprises offset, it causes a DC component to appear at the output of the A/D converter. Because not only the offset but also the DC content of the input signal appears as a DC component at the output, the offset cannot be distinguished from a DC signal. This is also shown in figure 4.11.

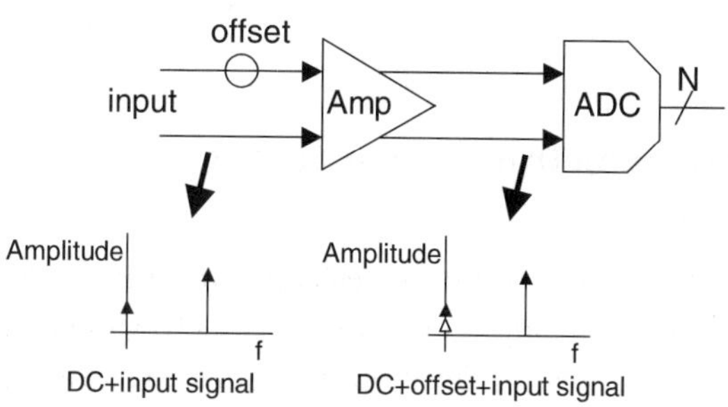

Figure 4.11: *Amplifier with offset, connected to an A/D converter, with input and output signal spectrum*

Chopping of the residue amplifier inputs is applied to be able to distinguish the residue amplifier offset from a DC signal component. The input of the amplifier is chopped as shown in figure 4.12. This means that the differential input of the amplifier is periodically reversed in polarity. Because the offset is added after the chopper, the DC component at the output of the amplifier only consists of the offset, while the DC component of the signal is located at the chop frequency. At the output of the amplifier the signal has to be chopped again to retrieve the original signal. This is shown in figure 4.12.

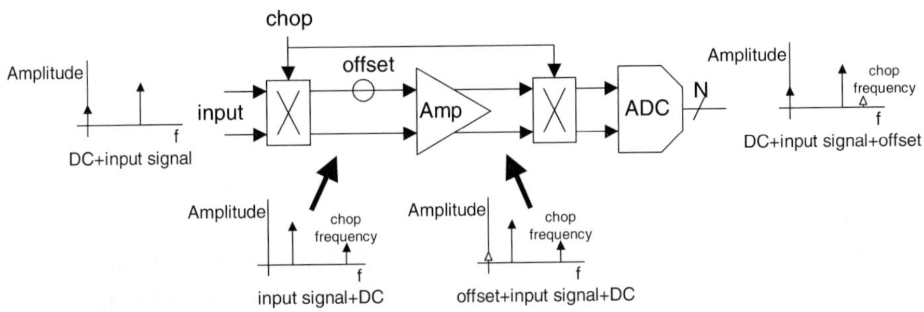

Figure 4.12: *Amplifier with choppers in order to move offset to distinguish offset from DC*

The chop state is not allowed to change just before the A/D converter quantizes the signal, since this puts high demands on the bandwidth of the system. The chop signal is therefore synchronized with the A/D converter clock signal, which means that the chop state does not change until just after the A/D converter has quantized the signal. This relaxes the additional demands on the bandwidth. Being able to change the chop state at the A/D converter clock edges, while preserving a 50% duty cycle, limits the possible chop frequencies to $\frac{f_s}{2}$, $\frac{f_s}{4}$, $\frac{f_s}{6}$, ..., with f_s as the A/D converter sample frequency.

4.5.2 Offset extraction from digital output

In the case of figure 4.12 the offset is perfectly separated from the DC content of the signal. In order to be able to detect the offset of the residue amplifier the digital output data is chopped again with the same chop signal as is used at the input of the amplifier, as shown in figure 4.13.

The resulting signal contains the offset component at DC. This signal is integrated in order to detect the sign of the offset. A positive offset causes the integrator content to be positive, while a negative offset results in a negative content. The input signal is averaged out during integration, since integration can also be referred to as a low pass filter. However, when the input signal frequency is close to the chop frequency, the input signal component is close to the offset component after chopping the digital output of the A/D converter. Figure 4.14 shows this in the frequency domain.

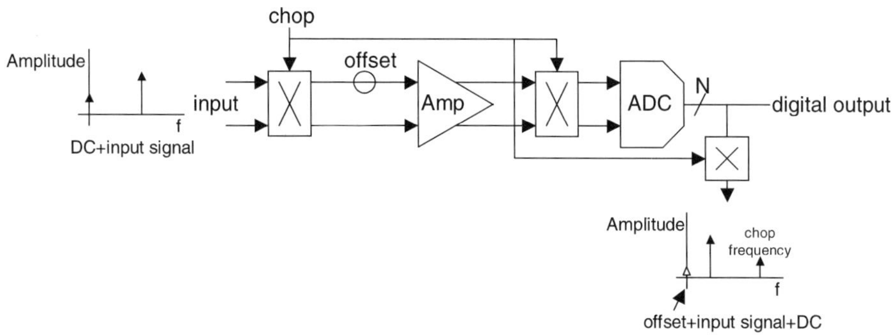

Figure 4.13: *Conversion system with amplifier choppers and a digital chopper for offset extraction*

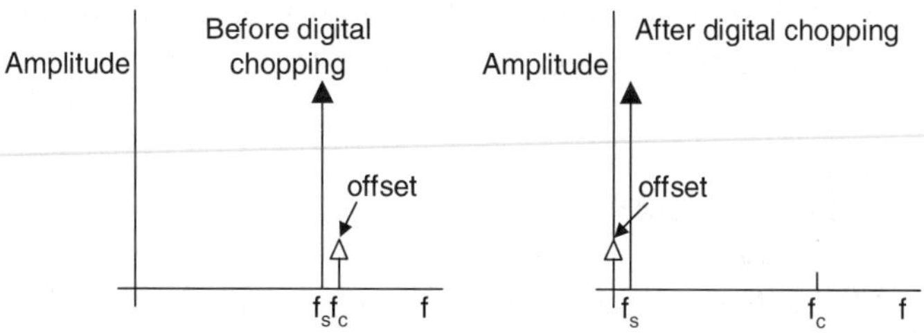

Figure 4.14: *Chopping of the amplifier input when the input signal is close to the chop frequency*

Integration of such an input signal also causes the integrator content to change, since the signal attenuation at low frequencies is small. This interferes with the detection of the offset sign. The minimum allowed ratio between the input signal frequency and chop frequency, for proper offset sign detection, can be calculated. This ratio between the input signal frequency f_{in}, sample frequency f_s and the chop frequency f_c can be defined as:

$$a = \frac{f_{in} - f_c}{f_s} \tag{4.5}$$

In the following derivation it is given that f_c is one of: $\frac{f_s}{2}, \frac{f_s}{4}, \frac{f_s}{6}, \ldots$. If the resolution of the digital signal is N bit, then a full scale sine wave, with frequency $f_{in} = a \cdot f_s + f_c$, the digital signal after chopping is equal to $2^{N-1} \sin(2\pi an)$, with n as the n-th sample. Integration of this signal yields:

$$M_{signal}(n) = \int_0^n 2^{N-1} \sin(2\pi ax)\, dx = 2^{N-1}\left(\frac{1}{2a\pi} - \frac{\cos(2\pi an)}{2a\pi}\right) \quad (4.6)$$

Suppose the smallest offset which has to be detected is $\pm\frac{LSB}{4}$ (appendix A), where $LSB = \frac{1}{2^N}$ of the A/D converter range, then the integrator content M due to offset is given by:

$$M_{offset,min}(n) = n \cdot \pm\frac{LSB}{4} = n \cdot \pm\frac{1}{4 \cdot 2^N}, \quad (4.7)$$

with the sign depending on the sign of the offset. Figure 4.15 shows the integrator content M as a function of the number of samples n of the minimum offset to be detected $-n \cdot \frac{1}{4 \cdot 2^N}$ and of the integrator content $M_{signal}(n) - n \cdot \frac{1}{4 \cdot 2^N}$ for several values of a with $N = 4$.

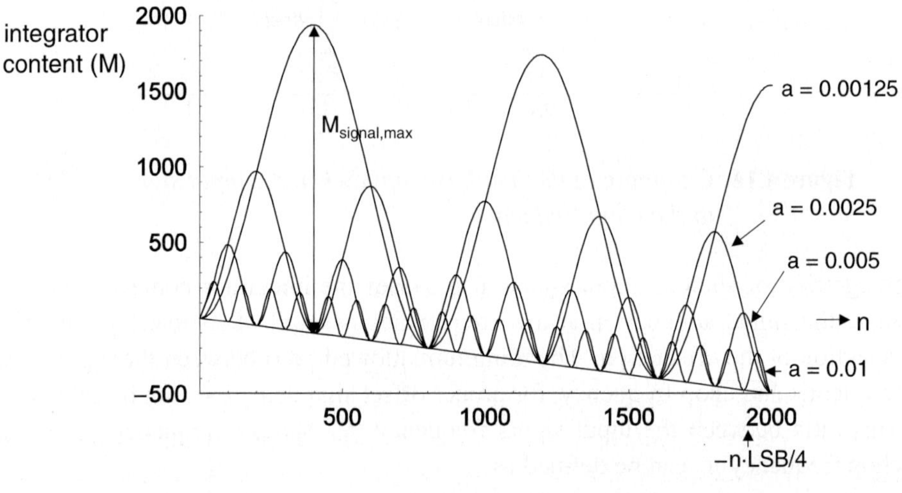

Figure 4.15: *Integrator content M for several values of 'a' with N = 4*

In figure 4.15 it can be seen that for a certain value of a, a wrong decision can be taken on the offset sign. For example, when $a = 0.00125$ and at $n = 1000$, the integrator content M is positive, even in the presence of a negative offset. This can be overcome when the number of integrated samples n is sufficiently large to enable a reliable decision to be made about the sign of the offset. The maximum value that the integrator content can reach is at the top of the signals (the integrator contents) shown in figure 4.15. This maximum value is calculated by equation 4.6 when n equals $|\frac{1}{2a} + \frac{m}{a}|$, with m as an integer. In this case the integrator content given by equation 4.6 becomes:

$$M_{signal,max} = \left| \frac{2^{N-1}}{a\pi} \right| \tag{4.8}$$

From this equation the minimum required number of integrated samples n can be calculated as a function of the ratio a, since the integrator content due to signal contribution always has to be smaller than the integrated offset:

$$M_{offset,min} > Int_{signal,max} \Rightarrow n > \left| \frac{4 \cdot 2^{2N-1}}{a\pi} \right| \tag{4.9}$$

This equation means that if a full scale input signal has a frequency between $|f_c - a \cdot f_s|$ and f_c, the integrator content after n samples is dominated by the input signal and not by offset. In this case the decision about the sign of the offset is not reliable. Equation 4.9 can also be written as a function of n:

$$|a| > \frac{4 \cdot 2^{2N-1}}{\frac{V_{pp}}{V_{fullscale}} \cdot n\pi} \tag{4.10}$$

The additional factor $\frac{V_{pp}}{V_{fullscale}}$ includes the effect when the peak-to-peak value of the input signal (V_{pp}) sine is smaller than the full scale value. For example, when the input signal is half of the maximum input signal swing, the value of a is halved. This means that input signal frequency is allowed to be twice as close to the chop frequency with respect to the frequency of a full range input signal. From equation 4.10 the 'forbidden' signal frequencies can be derived as a function of the signal amplitude and the integration period. This is shown in figure 4.16 for a chopping frequency of $f_c = \frac{f_s}{2}$.

For systems which are over-sampled this is in general not a problem, since most of these systems use anti-aliasing filtering. Due to this filtering there is no signal component left close to the Nyquist frequency. a can therefore be designed to be sufficiently large. However, to be able to deal with signals (including interfering signals), that are close to the chopping frequency, the result shown in figure 4.16 is not desirable because it cannot be guaranteed that there are no signals present

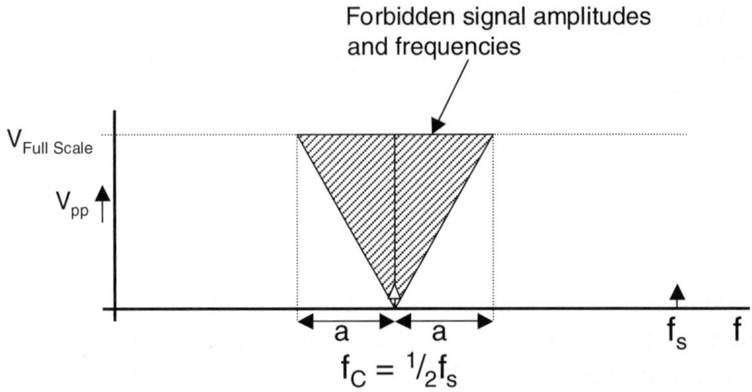

Figure 4.16: *The forbidden area in which no signal is allowed for proper offset detection for* $f_c = \frac{f_s}{2}$

in the 'forbidden' areas. The size of the 'forbidden' areas can be reduced by using a chopping signal with different frequencies. This is discussed in the next section.

4.5.3 Pseudo random chopping

If instead of a single chopping frequency a chopping signal with more frequency components is used, the size of the forbidden area shown in figure 4.16 is reduced, since the input signal is spread over more frequencies by chopping. This means that the signal power at each chopping frequency component is less than the signal power of the original signal. In the voltage spectrum, the signal is divided equally over all the chopping frequency components with a value in each peak of:

$$V_{amp,choppeak} = \frac{V_{pp}}{\sqrt{N_{peak}}}, \tag{4.11}$$

where $V_{amp,choppeak}$ and V_{pp} are the signal amplitude at each chop frequency and the original signal respectively. The chop signal consists of N_{peak} frequency components. The effect of multiple chopping frequencies is shown in figure 4.17 for a case with four chopping frequencies.

Note that a in figure 4.17 is smaller than a in figure 4.16. When the input signal is located in one of the forbidden areas, the component folded back to (close to)

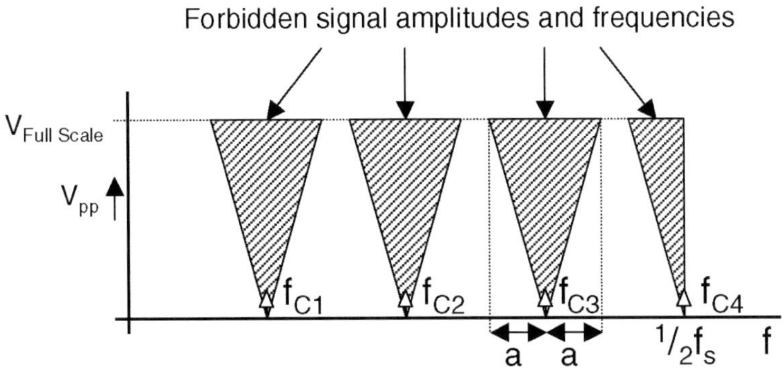

Figure 4.17: *The forbidden areas in which no signal is allowed for proper offset detection*

DC is sufficient to disturb the offset detection. As can be seen in figure 4.17, when the input signal frequency equals exactly one of the chopping frequencies, the integrator handles this as an offset, since part of the input signal appears at DC. To be sure that this component is always less than the minimum offset which has to be detected, the input signal has to be spread through the spectrum with no components larger than the minimum detectable offset. This makes demands on the number of chopping frequencies. When the input signal is distributed equally over N_{peak} frequencies, the signal amplitude at each of these frequencies i is then given by:

$$V_{amp,signal_i} = \frac{V_{pp}}{\sqrt{N_{peak}}} \qquad (4.12)$$

This means that when the input signal frequency falls together with one of the chop frequencies (and has the same phase), the amplitude of the component which will appear at DC is given by equation 4.12. This amplitude has to be smaller than the minimum offset which has to be detected. For a full scale input signal the minimum number of frequencies N_{peak} is then calculated by:

$$N_{peak} \geq \frac{2^{2(N+1)}}{2} = 2^{2N+1} \qquad (4.13)$$

The minimum offset is set at $\frac{LSB}{4}$. The division by 2 in equation 4.13 is the factor between the full scale and the amplitude of the input signal.

To generate a chop signal with sufficient frequency components, a digital pseudo random-noise generator can be used [114]. An example is shown in figure 4.18.

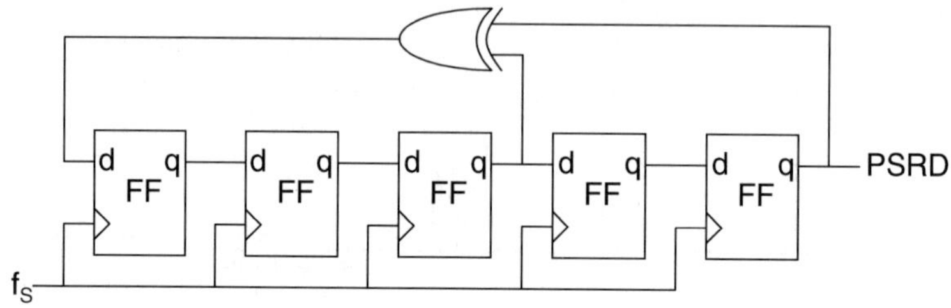

Figure 4.18: *Pseudo random-signal generator with five delay elements*

The number of delay elements used in the pseudo random generator determines the length of the pseudo random cycle. In the example of figure 4.18 the number of frequencies of the generated signal is 15. The amplitude of each component is $\frac{1}{\sqrt{15}}$ or –11,8 dB with respect to a single tone signal, as is shown in figure 4.19.

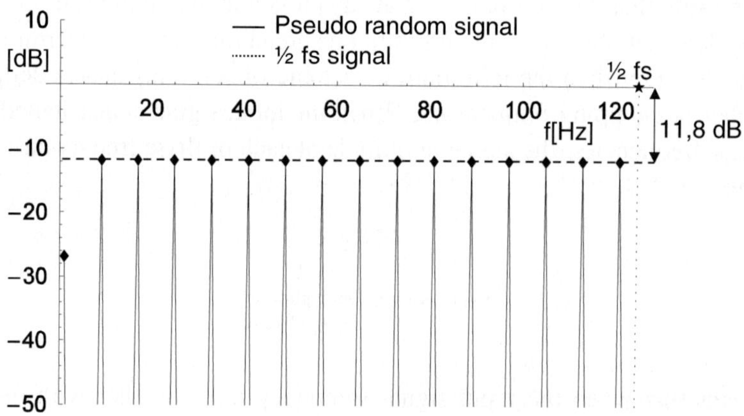

Figure 4.19: *Frequency spectrum of an $\frac{f_s}{2}$ signal and the pseudo random signal generated from figure 4.18*

A pseudo random signal consists of a random sequence of ones and zeros. After a full pseudo random cycle the result of the integration of this signal is zero, since the number of ones and zeros in a full cycle is (almost) equal. The minimum required integration length for proper offset detection is therefore equal to the pseudo random cycle length. The minimum required pseudo random cycle length is derived by calculating the minimum required number of peaks (N_{peak} from equation 4.13).

4.5.4 Offset extraction and analog compensation

After an integration cycle the sign of the offset can be derived from the integrator content. A positive integrator content is the result of a positive offset. This result can be used to calibrate the offset of the amplifier from figure 4.12[2]. By using a D/A converter, the information from the digital domain can be converted into a signal which compensates the amplifier offset. This total calibration loop is shown in figure 4.20.

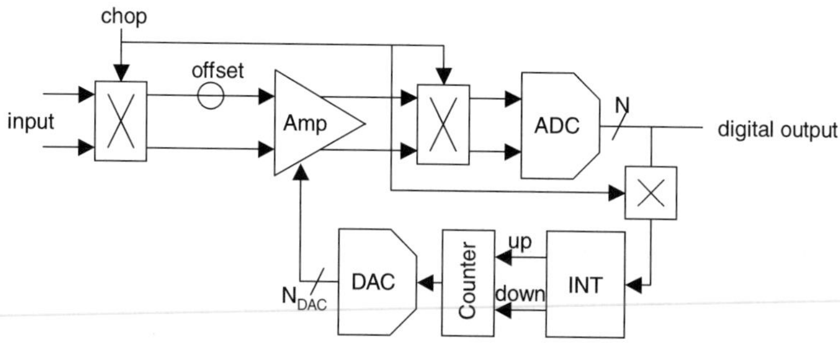

Figure 4.20: *Total amplifier offset calibration loop with digital offset extraction and analog compensation*

For the A/D converter shown in figure 4.20 it makes no difference whether the input signal is chopped or not, besides a Nyquist signal bandwidth requirement. However, in a sub A/D converter (i.e. the fine A/D converter in a two-step architecture) the signal bandwidth already has to fulfil this requirement. The second analog chopper just at the input of the A/D converter can therefore also be moved to the output of the A/D converter. In this case the chopper becomes a simple digital chopper and the digital chopper in front of the digital integrator shown in figure 4.20 can be removed since the output of the (sub) A/D converter can be integrated directly. This results in the calibration loop of figure 4.21.

A counter is used to drive the offset-compensating D/A converter. When a positive offset is detected, the integrator block gives a down pulse to the counter and the D/A converter value is decremented. The integrator content is set to zero and a new integration cycle starts. This integrator reset cuts the loop and no instability problems can therefore occur. When the offset is sufficiently small, less than a certain threshold, the integrator gives no updates to the counter. This is to circumvent the offset to toggle around zero. This would give a frequency component at

[2] A similar offset detection method has been published later than [58] in [115].

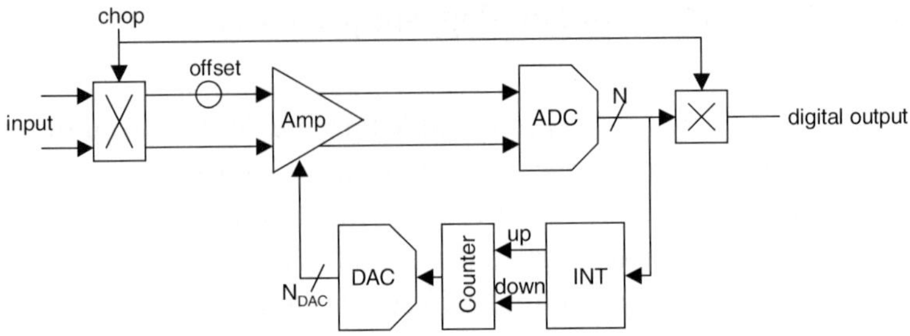

Figure 4.21: *Total amplifier offset calibration loop with improved chopping*

the integration cycle length. This threshold can be calculated by multiplying the minimum offset which has to be detected, for example $\frac{LSB}{4}$, by the total number of integrated samples. Figure 4.22 illustrates the operation of the calibration loop.

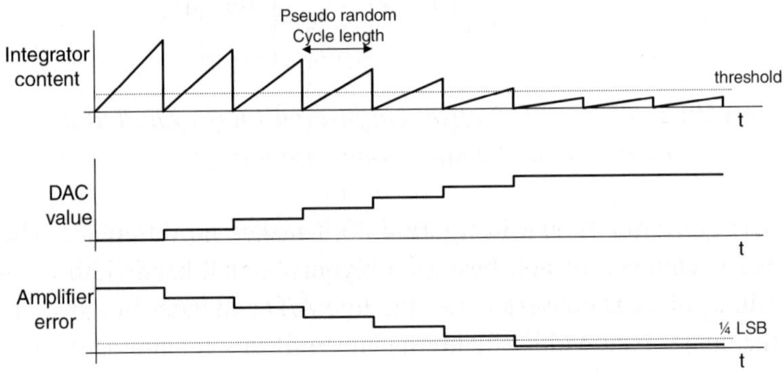

Figure 4.22: *Integrator and D/A converter signals of the calibration loop, together with the resulting error*

The demands on the D/A converter are determined by the step size of the compensation. If the integrator is able to detect offsets of $\frac{LSB}{4}$ the step size (or resolution) of the D/A converter should be less or equal to that value. The linearity of the offset compensation D/A converter is not critical since the D/A converter is part of the regulation loop.

The described calibration loop enables offset calibration of the amplifier during normal operation. The amplifier and A/D converter are transparent for normal input signals, while the offset of the amplifier is extracted in the digital domain. This enables the use of a filter or integrator with a large time constant, without the need for accurate or sensitive analog components. The extracted offset is used to drive a D/A converter to compensate for the amplifier offset.

4.5.5 Offset extraction in a dual-residue two-step converter

This calibration algorithm is not only applicable for the offset of a single amplifier in a single residue system. When dual-residue signal processing is used, as described in section 4.3.2 and shown in figure 4.23a, the effect of offset on the residue amplifiers on the transfer curve of the sub A/D converter is shown in figure 4.23b. The offset of the residue amplifiers can be made small by design, but this requires large input devices. To reduce the input capacitance of the residue amplifiers, the offsets of both amplifiers are calibrated.

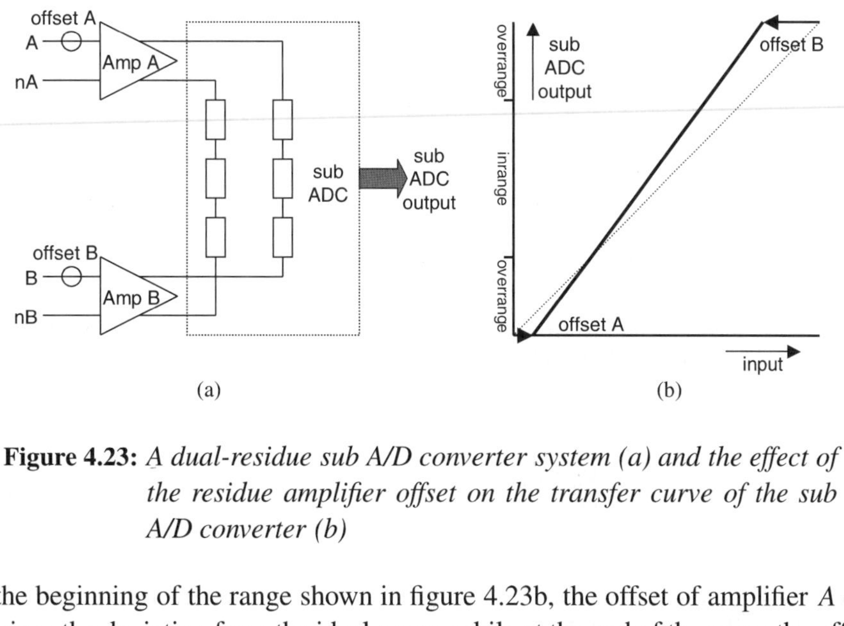

Figure 4.23: *A dual-residue sub A/D converter system (a) and the effect of the residue amplifier offset on the transfer curve of the sub A/D converter (b)*

At the beginning of the range shown in figure 4.23b, the offset of amplifier A determines the deviation from the ideal curve, while at the end of the range the offset of amplifier B determines the deviation from the ideal curve. The offsets of both amplifier A and amplifier B, as shown in figure 4.24a, are split up into a common component and a differential component because they require a different detection and calibration scheme, as will be explained later. The common offset component has equal values and equal signs for both amplifiers (figure 4.24b) and the differential offset component has equal values but has opposite signs (figure 4.24c).

The offsets of both amplifiers can be written as a function of the common and differential offset components:

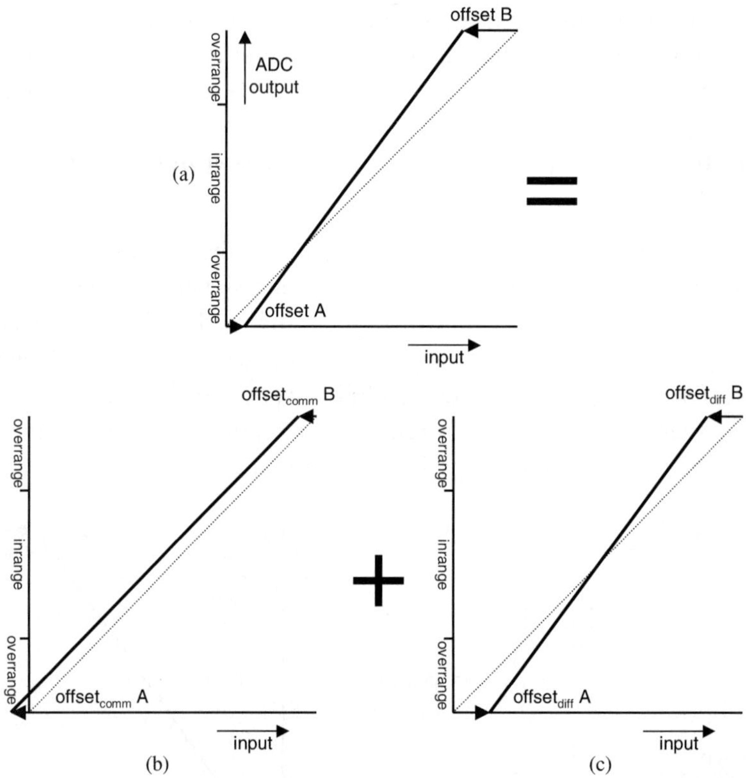

Figure 4.24: *The offsets of amplifiers A and B (a) split up into common (b) and differential (c) components*

$$\begin{aligned} V_{offset\ A} &= V_{offset_{comm}} + V_{offset_{diff}} \\ V_{offset\ B} &= V_{offset_{comm}} - V_{offset_{diff}} \end{aligned} \tag{4.14}$$

Equation 4.14 can be rewritten to calculate the common and the differential offset component from the amplifier offsets:

$$\begin{aligned} V_{offset_{comm}} &= \frac{V_{offset\ A}+V_{offset\ B}}{2} \\ V_{offset_{diff}} &= \frac{V_{offset\ A}-V_{offset\ B}}{2} \end{aligned} \tag{4.15}$$

The effect of the common offset component is a DC shift of the transfer curve of the A/D converter. This does not reduce the non-linearity of the two-step A/D converter. In principle this offset does not have to be compensated for. However, since the common offset component causes a shift of the residue signal in the fine A/D converter range, it uses part of the over-range of the fine converter while this is intended for the coarse A/D errors. When the common offset component is reduced as well, the total fine A/D converter over-range can be used for the coarse A/D converter errors. This allows the power of the coarse A/D converter to be reduced to the minimum. The differential offset component, however, directly reduces the non-linearity of the two-step A/D converter. Both the common and differential offsets therefore need to be calibrated.

For the sake of simplicity, the offset detection without chopping of the residue amplifier inputs is considered first. In order to be able to extract the sign of the common or differential amplifier offset component, the effect of the respective offset component should be a signal at DC. The sign of the offset can then be detected by integration of the digital data (which is actually low-pass filtering). This is illustrated in figure 4.25.

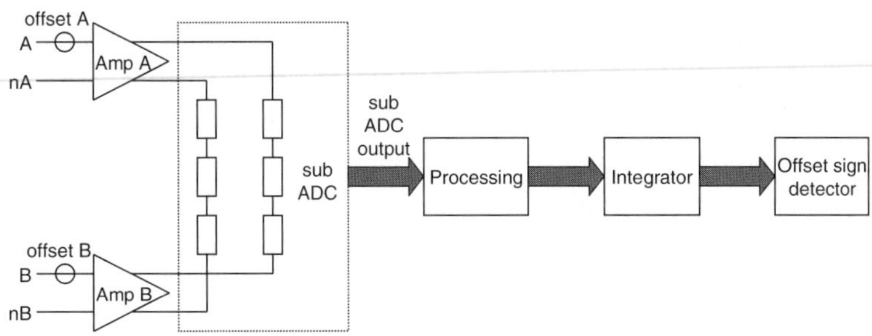

Figure 4.25: *Offset sign detection by integration of the digital output of a dual residue amplifier conversion system*

The additional processing in the digital signal path is to transform the signal such that a signal at DC appears when there is amplifier offset. It is clear that in this case a DC component in the input signal will interfere with the offset detection, although chopping of the residue amplifiers (described in section 4.5.1) removes this restriction, as is explained later.

Common offset

First the common offset component from figure 4.24b is considered. In this case the additional signal processing shown in figure 4.25 only puts the zero reference in the middle of the output range, as is shown in figure 4.26a.

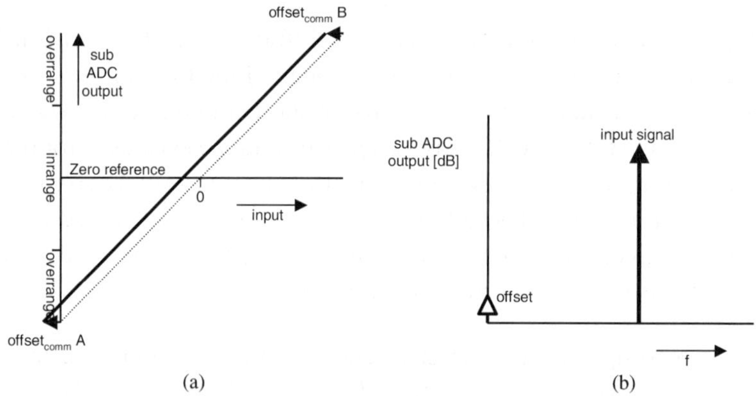

Figure 4.26: *Input of the integrator (a), with the zero reference in the middle of the range and the spectrum (b) of the sub A/D converter output with only common offset*

When the input signal of the residue amplifiers is a (randomly) varying signal (for example a sine with zero DC component), the signal varies at the x-axis of figure 4.26a. When both amplifiers of figure 4.25 have an equal offset, a DC component appears in the output signal (figure 4.26b), which is caused by this offset. When the output signal of the sub ADC is applied to the integrator from figure 4.25, this DC component leads to a positive integrator content. This is a measure for the sign of the offset and can be used in a calibration loop to remove the offset.

Differential offset

When the offsets of both residue amplifiers from figure 4.25 consist of only a differential component the transfer curve of the sub A/D converter is shown in figure 4.24c. When the same processing as described above with common offset is used, the input of the integrator as a function of the input signal is shown in figure 4.27a.

When in this case the input signal of the residue amplifiers is a (randomly) varying signal (for example a sine with zero DC component) similar to that described above, there is no offset component visible at DC. Figure 4.27b shows that the error is only visible as a gain error, but this cannot be detected with the low-pass filtering of an integrator. To enable the detection of the differential offset as a DC component a larger part of the total A/D converter range (more sub-ranges) is

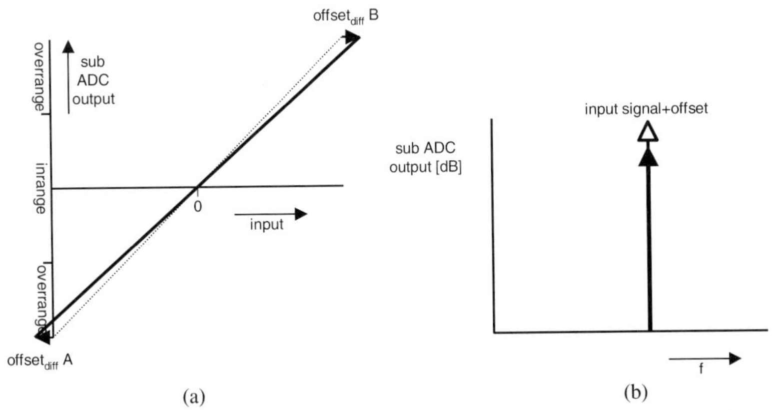

Figure 4.27: *Input of the integrator (a), with the zero reference in the middle of the range and the spectrum (b) of the sub A/D converter output with only differential offset*

considered, as shown in figure 4.28a. The spectrum of the A/D converter output is shown in figure 4.28b, when an example input signal without a DC component is applied.

The differential offset is spread through the spectrum and indicated as 'white noise' in figure 4.28b. In reality, however, it shows up as discrete peaks, which are input-signal-frequency-dependent. The additional digital processing detects whether the output signal passes a threshold which lies in the middle of a subrange. According to this detection, the proper value is subtracted from the digital code. The result is shown in figure 4.28c. This additional processing works in a similar way to the two-step processing, where the threshold lies at a subrange transition. If the applied input signal contains no DC, the differential offset can still not be detected by low-pass filtering, since there is no offset component at DC. By inverting the digital code in the odd 'subranges' as shown in figure 4.28d, part of the differential offset appears at DC in the spectrum (figure 4.28e). Due to a differential offset the transfer curve is always above (or equal), or always below (or equal) to the ideal transfer curve (depending on the sign of the differential offset). This is illustrated in figure 4.28d. When any (random) signal without a DC component is applied to the A/D converter, which means that the input signal varies over the x-axis of figure 4.28a, c and d, part of the differential offset appears at DC. By low-pass filtering or digital integration, the sign of this offset can be detected, in a similar way to the common offset.

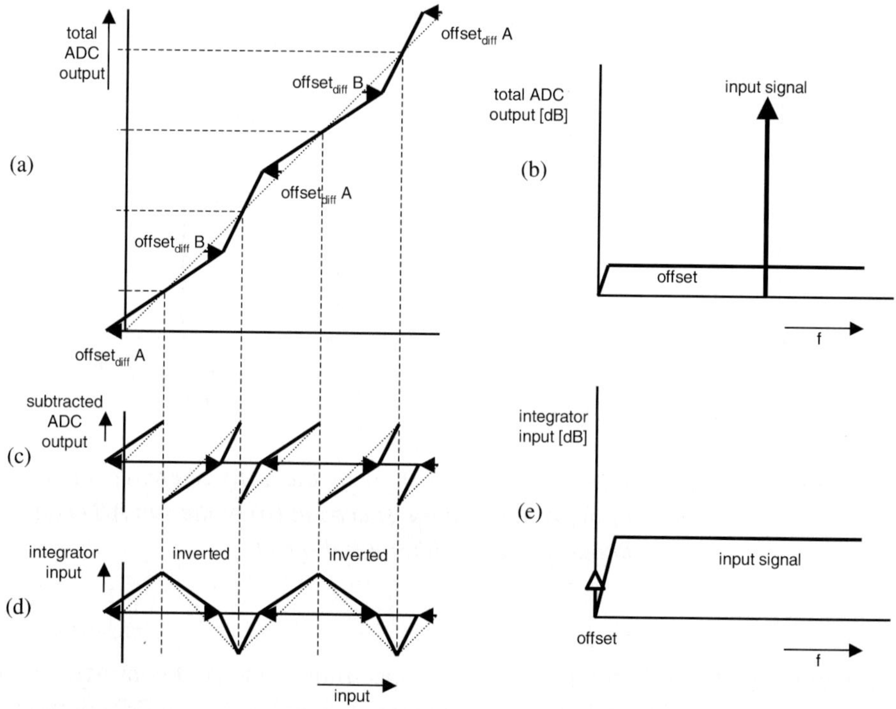

Figure 4.28: *Output of the total A/D converter with 4 subranges, only differential offset on the residue amplifiers (a) and the spectrum of an example signal (b), subrange processing (c) and partial inverted integrator input (d) with the resulting spectrum (e)*

Chopping

To be able to distinguish the offset error from the (DC) input signal, the inputs of the residue amplifiers are chopped, as explained in section 4.5. Since two amplifiers are chopped, there are four connection possibilities. With one un-chopped connection, this results in three different chopping methods. This is illustrated in figure 4.29.

Chopping method 1 changes the polarity of the inputs of both amplifiers. Chopping method 2 also changes the polarity of the inputs and interchanges the inputs of amplifier A and amplifier B. The third chopping method leaves the polarity the same and only interchanges the inputs of amplifier A and amplifier B. The three chopping methods deal with the common and differential offset components

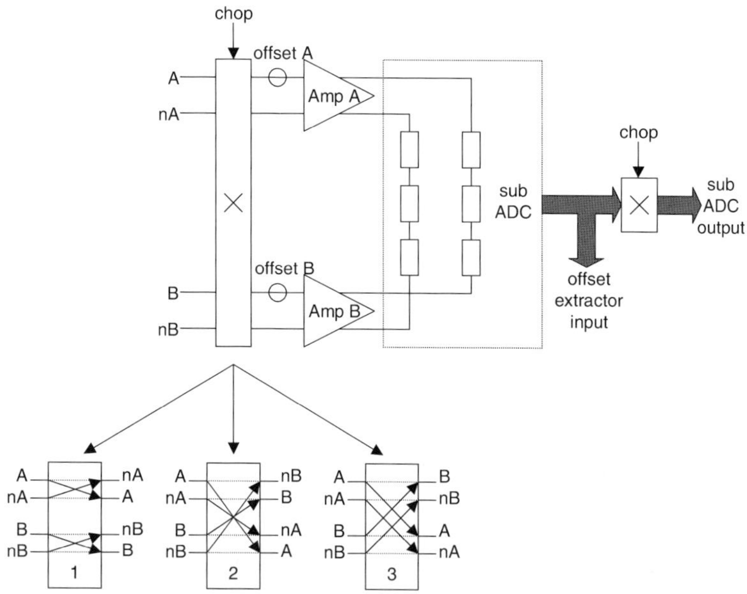

Figure 4.29: *Chopping of dual-residue amplifiers with three possibilities of chopping*

in a different way. Only the first chopping method is considered here, since both common and differential offsets can be detected using this chopping method. The other two chopping methods can be used to detect only common or only differential offset, but this is not discussed here.

When the inputs of the residue amplifiers are chopped (with chopping method 1) under the condition that only common offset is present at the residue amplifiers, the transfer curves of both chop states are shown in figure 4.30. The chop state $nochop$ has the same transfer curve, as shown in figure 4.26a. The signal which is applied to the input of the integrator shown in figure 4.25 comes from one of the transfer curves of figure 4.30, depending on the chop state.

The first effect of chopping is that the signal component is inverted for an equal sub A/D converter input signal. The second effect of chopping is that the offset component does not change sign for an equal sub A/D converter input signal. This is clear from figure 4.29 since the amplifier offset is added after chopping of the signals. Suppose the example signal V_x shown in figure 4.30 is the input signal of the sub A/D converter in figure 4.25. The value on the $nochop$ or $chop1$ curve is then applied to the integrator of figure 4.25, depending on the chop state. The

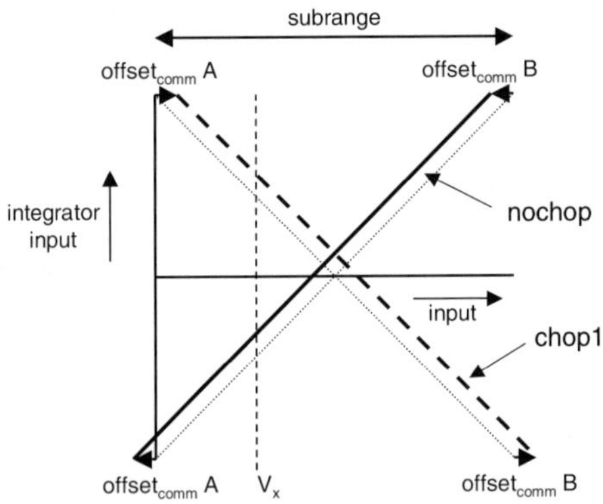

Figure 4.30: *The transfer curves of both chop states at the integrator input with only common offset*

integration causes the signal component to be canceled, while the offset component results in a positive integrator content. This indicates the sign of the offset. Furthermore, with a 'randomly' varying input signal at the x-axis and a (pseudo-random) chop signal, the input signal is canceled after integration, in much the same way as described in section 4.5.2.

When the inputs of the residue amplifiers are chopped (with chopping method 1) under the condition that only differential offset is present at the residue amplifiers, the transfer curves of both chop states are shown in figure 4.31. The chop state *nochop* has the same transfer curve as shown in figure 4.28d. The signal which is applied to the input of the integrator shown in figure 4.25 comes from one of the transfer curves of figure 4.31, depending on the chop state.

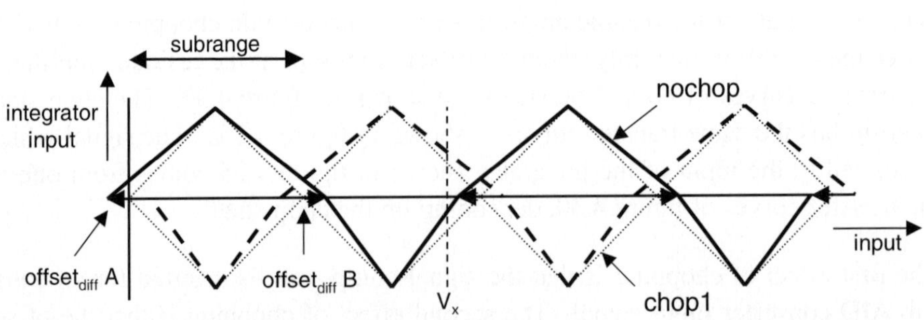

Figure 4.31: *The transfer curves of both chop states at the integrator input with only differential offset*

The same effect of chopping applies for differential offset as for common offset. The signal component is inverted, while the offset component does not change sign. When an example signal is applied (V_x in figure 4.31), depending on the chop state, the value on the *nochop* or *chop*1 curve is applied to the integrator of figure 4.25. The integrator causes the signal component to be canceled, in a similar way to that described above for a common offset component, while the offset component causes the integrator content to be positive. This is caused by the fact that the transfer curve with differential offset is always above (or equal to) the curve without offset in figure 4.31. The positive integrator content is a measure for the sign of the offset. Also with a 'randomly' varying input signal at the x-axis and a (pseudo-random) chop signal, the input signal is canceled due to integration, in a similar way to that described in section 4.5.

The additional signal processing for both component extractions is different, with the result that separate paths are used for the detection of the common and differential components. The complete system for extracting both the common and differential offset in the digital domain and compensating the extracted offset in the analog domain is shown in figure 4.32.

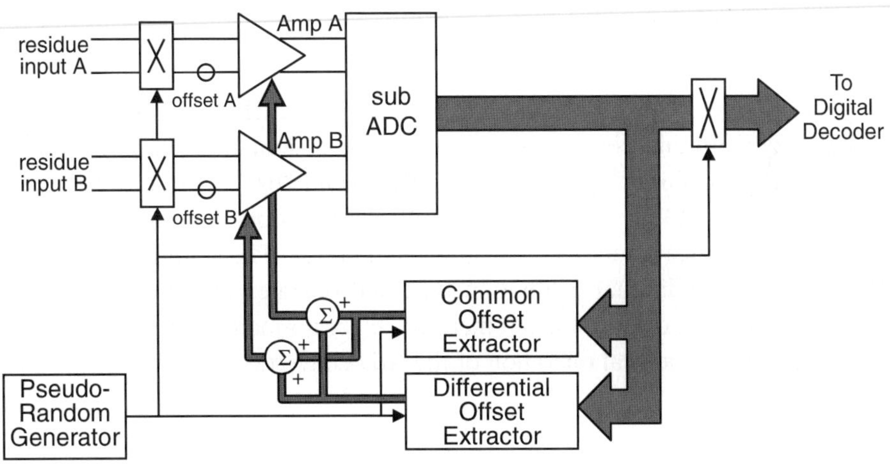

Figure 4.32: *Complete digital offset extraction and analog compensation system*

The effect of the difference of the compensation of the common offset and of the differential offset is illustrated by the signs at the summation points of the compensation D/A converter signals. A common offset results in an equal change of both compensation D/A converters, while a differential offset results in a change with opposite sign.

Because of the additional processing required to detect the differential offset, a common offset component is averaged out in the differential offset extractor. Visa versa, the same holds for a differential offset component. A differential offset component is averaged out in the common offset extractor. The common and differential offset components can therefore be extracted simultaneously without interfering with each other. A minor effect occurs when the input signal is a DC signal or an input signal with a frequency which is lower than the cycle length of the calibration. In this case it is not possible to distinguish between a common and differential offset component. To overcome this problem a detector is used to check if the input signal is distributed sufficiently over the whole subrange. In the digital domain, the subrange is divided into 4 bins. The number of occurrences within each bin is counted. When the number of occurrences in all of the 4 bins within an integration period is above a certain threshold, the signal is then distributed sufficiently. If this is not the case the outcome of the integration cycle is not used.

4.5.6 Conclusions

The mixed-signal chopping and calibration technique enables offset extraction in the digital domain and compensation in the analog domain. The powerful signal processing capability of the digital CMOS technology is used to make large time constants, enabling highly accurate offset extraction. The required analog circuitry is not critical in terms of accuracy and power. The calibration technique does not require additional calibration time and operates under normal conversion conditions. The calibration can therefore be performed continuously without interfering with the normal operation of the A/D converter.

Chapter 5

A 10-bit two-step ADC with analog online calibration

5.1 Introduction

The scope of the experiment [28] described in this chapter is to verify the dual-residue signal processing [3] from section 4.3.2 in combination with analog offset calibration [106] in the two-step architecture [26] (section 3.4). In the previous chapters the basic architectural choices that involve both the accuracy of the converter and the speed were discussed. The experiment described in this chapter shows an implementation based on the theory developed in the previous sections. The quantization of video signals is used as a vehicle. Application as an embedded video converter requires that the specifications in table 5.1 are met.

Table 5.1 shows the general requirements for video conversion for baseband signals. Cost requirements demand that the A/D converter is embedded in a digital CMOS chip. The technology, supply and sample rate choices are therefore given by the video architecture [116]. In video systems, the signal from the tuner is down-converted in the IF section. Subsequently, the CVBS signal is split into chrominance and luminance signals (YUV). The choice is between a single high-performance A/D converter for CVBS, which is what is focused on in this experiment, or a triple A/D converter for YUV. In the A/D converter for CVBS discussed here, the color signal is still at 4.433 MHz, which means that the A/D converter performance at this frequency will determine the quality of the color performance.

Table 5.1: *Requirements for video conversion*

Technology	Digital CMOS without options
Resolution	8 bit for YUV, 10 bit for CVBS
Supply	Single supply acc. to architecture
Color carrier	<1° phase delay at 4.433 MHz
Effective bandwidth	3 MHz for YUV, 5.5 MHz for CVBS
Sample Rate	13.5, 27, 54 MSample/s depending on DSP choice
DNL	Strict monotonical (<0.7 LSB)
INL	Up to 3 LSB tolerable
Power dissipation	<250 mW (for embedding as low as possible)
Area	<1 mm^2

The DNL is the most critical parameter for luminance. In scenes with low luminance variation a large DNL will result in visible artifacts ('contouring'). In CVBS this demand is more stringent as the dynamic range of the signal is larger. The dynamic properties of the A/D converter also need to be sufficient for performance at the color carrier frequency. INL errors translate into distortion at higher signal frequencies, thereby deteriorating the color performance.

In order to enable embedding in larger processing chips, the power consumption and the area must be as low as possible. The target of 250 mW was chosen, which is a quarter of the maximum power dissipation in a low-cost IC-package.

The traditional two-step architecture requires an amplifier stage with an exact gain factor. Using dual-residue signal processing, the gain requirements are limited to the gain matching of two residue amplifiers. The offset on these residue amplifiers makes a major contribution to the INL of the converter. Their offset is calibrated in order to reduce the input capacitance of these amplifiers. Due to this offset calibration, the total intrinsic capacitance can be reduced significantly, thereby reducing the power. This calibration is relatively simple since only two components need to be calibrated. The performance improvements due to dual-residue signal processing with offset calibration result predominantly in better linearity metrics such as DNL, INL and spurious-free dynamic range (SFDR).

The accuracy of the two-step converter depends on the subtracting D/A converter accuracy, the coarse and fine quantizer accuracy and the summing accuracy, all of which are very much dependent on the actual circuit implementation. The accuracy of the implemented converter described in this chapter depends mainly on the ladder resistor matching and the offset of just two residue signal amplifiers.

Section 5.2 discusses the two-step architecture. In this section, the architectural solutions for problems which can arise in a two-step A/D converter are introduced. Section 5.3 explains the design of critical circuits of the A/D converter. Section 5.4 shows the experimental results of this design. The discussion and the conclusions are presented in section 5.5 and section 5.6.

5.2 Two-step architecture

A detailed block diagram of the two-step A/D converter is shown in figure 5.1.

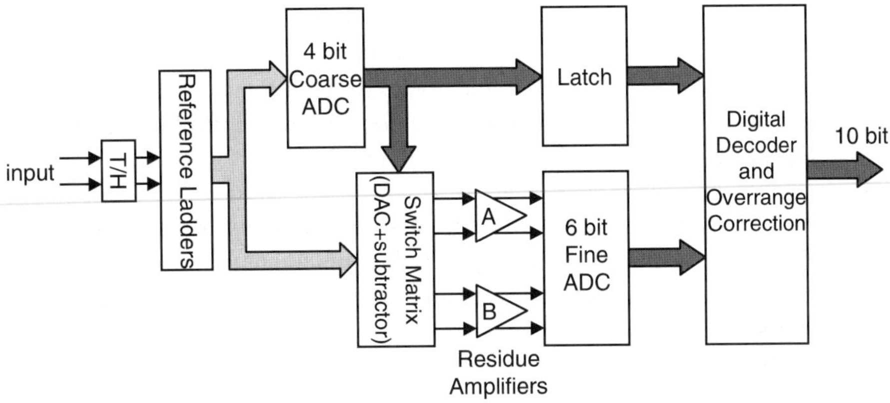

Figure 5.1: *Block diagram of the two-step A/D converter*

The differential analog input signal is sampled with a pseudo-differential track-and-hold. After the track-and-hold, the differential signal is applied to two floating resistor ladders via buffers. These ladders must have 10-bit accuracy with respect to resistor matching and settling at the end of the hold period. 17 coarse comparators, which perform the coarse quantization, are connected to this ladder. Since the resistor ladders move with the input signal, the common-mode level of each comparator input signal is the same. This also holds for the switches from the switch matrix. These switches perform the D/A converter and subtractor function in order to construct the pairs of differential residue signals according to the coarse decision. This will be explained in the following sections. Two differential amplifiers amplify these residue signals. The offsets on these amplifiers must be low because they determine the integral linearity of the A/D converter. After the residue signals are amplified, they are applied to the fine A/D converter, which performs the final quantization. The (digital) delay between the coarse A/D converter and the digital decoder is necessary because the coarse and fine quantizations do not take place at the same time. The digital decoder combines the outputs of the coarse and fine quantizer and corrects the coarse code for over-range detected in the fine quantizer.

The partitioning of coarse and fine bits is an important design parameter in the two-step architecture, as is shown in section 3.4. The input circuitry of both coarse and fine (flash) A/D converters only has to drive a reduced number of comparators compared to a full-flash A/D converter, resulting in a major capacitive load reduction. The gain between coarse and fine A/D converters reduces this load even further while also lowering the accuracy requirements of the fine comparators. This allows the two-step A/D converter to achieve a high sampling speed at low power consumption. This converter has 4 coarse bits, 6 fine bits and an amplifier gain of 10, which means that the accuracy required of the fine A/D converter is reduced by a factor of 10. A better value would be 5 coarse and 5 fine bits according to equation 3.18, but this A/D converter does not use a full bit over-range in the fine A/D converter, which changes the optimization.

The two-step architecture is suitable for high-speed sampling but the achievement of high linearity presents extra challenges to the implementation. In the following paragraphs the most important issues are analyzed and the solutions are presented.

5.2.1 Coarse quantizer accuracy

The coarse quantization determines in which subrange the input signal lies. An error in the coarse quantizer (due to comparator offset) results in residue signals which are out-of-range for the fine quantizer. From section 3.4 it follows that, without redundancy, the comparators of the coarse quantizer would need to have the same accuracy of the total A/D converter (N bits). Suppose an A/D converter with $N = 10$, 4 coarse bits and an input signal with an amplitude $V_{pp} = 1.6$ V. To achieve a yield of 99%, 4σ spread of the offset of the comparator input stages has to be smaller than a quarter of an LSB (derived using equation 3.1):

$$4\sigma_{V_{offset,coarse}} < \frac{1}{4} \cdot \frac{V_{pp}}{2^N} \Rightarrow \sigma_{V_{offset,coarse}} < 0.1mV \qquad (5.1)$$

Such low offsets can only be obtained with very large transistors, as described in section 2.2.1, with a capacitive load in the ten pF range because the areas of these transistors determine their matching [4]. This would have a negative effect on the sampling speed or power consumption of the A/D converter. To overcome the stringent offset requirement in the coarse quantizer, over-range is applied in the fine quantizer, as described in section 3.4 [29]. In this case the accuracy of the coarse comparators ($\sigma_{V_{offset,coarse}}$) is dependent on the number of over-range comparators ($\#ORC$) which are added to the fine quantizer:

$$4\sigma_{V_{offset,coarse}} < (\frac{1}{4} + \#ORC) \cdot \frac{V_{pp}}{2^N} \tag{5.2}$$

Equation 5.2 shows that the offset required in the coarse quantizer is relaxed as more over-range comparators are added to the fine quantizer. Because the addition of extra comparators to the fine quantizer will make a negative contribution to the speed and power consumption, the number of over-range comparators must be optimized for these parameters.

5.2.2 D/A converter and subtractor accuracy

When the coarse quantizer has decided in which subrange the input signal lies, the result is applied to the D/A converter which generates an analog signal proportional to the coarse decision. This signal is subtracted from the analog input signal, generating a residue signal that is applied to the fine quantizer. The accuracy of both the D/A converter and the subtractor is very important because it determines the quality of the residue signal and can therefore limit the performance of the A/D converter.

In this design, the references for the coarse A/D converter and the D/A converter are the same [117]. In this case the reference used to select the proper subrange and the reference which is subtracted from the input signal are equal. When this is not the case, a gain error between input signal and D/A converter output [88] may lead to non-monotonicity or missing codes. Possible errors in the subtractor include offset and distortion. The first can cause the residue signal to be out-of-range for the fine quantizer, and the second distorts the residue signal, which causes distortion in the transfer curve of the total A/D converter. To overcome these problems, the D/A converter and subtractor function are implemented as switches connected to the moving reference ladders. This is illustrated as a single residue system in figure 5.2 [90].

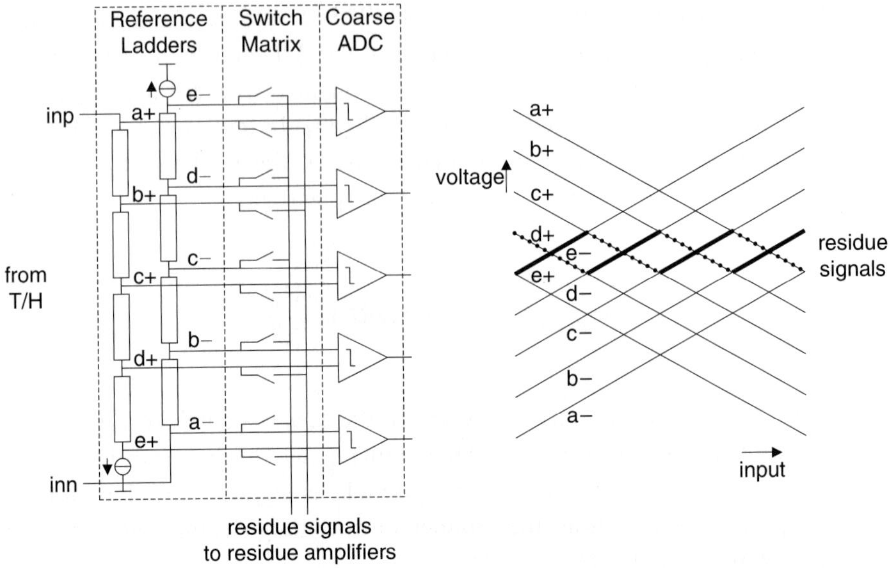

Figure 5.2: *Reference ladders, coarse A/D converter, D/A converter and subtractor generating the residue signals*

In order to generate the proper residue signals shown on the left in figure 5.2, the switches are set according to the coarse quantization. This means there are no components that contribute to offset or distortion present in the signal path up to the residue amplifiers. The coarse quantizer reference, the D/A converter reference and the subtractor are laid out as one compact block.

5.2.3 Coarse and fine A/D converter references

After the coarse quantization, the residue-signals have to be applied to the fine quantizer. A major problem for high accuracy is that the range of the fine quantizer has to fit perfectly in the selected subrange. If this is not the case, the fine codes will not be distributed equally over the subrange and missing codes can occur. Figure 4.4 illustrates the case where the range of the fine quantizer is larger than the coarse subrange. Dual residue signal processing (4.3.2) is used to circumvent the occurrence of large code jumps at the subrange transitions. This is implemented by connecting both the top and the bottom of the fine reference ladder to the coarse reference ladder as shown in figure 5.3. In this figure the over-range has been left out for the sake of simplicity.

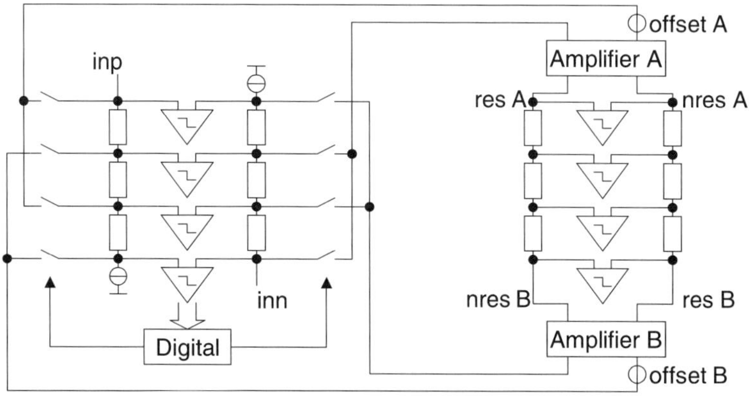

Figure 5.3: *Generation of four residue signals (ignoring over-range)*

This means the range of the fine quantizer reference adapts itself automatically to the coarse subrange. It also means that small differences between the resistor values of the coarse reference ladder will not lead to missing codes. Another advantage is achieved by switching in such a way that the same amplifiers, buffers and fine comparators are used when the change is made from one subrange into the adjacent subrange (section 4.3.2) [91]. In this case one side of the fine reference ladder remains connected to the coarse reference ladder while the other side changes its tapping point, as shown in figure 5.3. The effect of this improved switching on the overall transfer curve of the A/D converter is shown in figure 5.4.

Continuity and monotonicity of the converter are guaranteed, even with residue amplifier offset, due to this improved switching.

5.2.4 Amplifier gain and offset accuracy

The two pairs of differential residue signals, which have been generated by switching from the coarse reference ladder, are amplified before they are applied to the fine quantizer. This reduces the required accuracy of the fine quantizer. After a gain factor of 10, the LSB size is enlarged from 1.5 mV to 15 mV. This allows the size of the input transistors of the fine comparators to be reduced and therefore the speed to be increased or the power to be reduced. Because the 4 residue-signals consist of 2 differential signals, 2 differential amplifiers are used. The offsets of these amplifiers are important since they make a major contribution to the integral

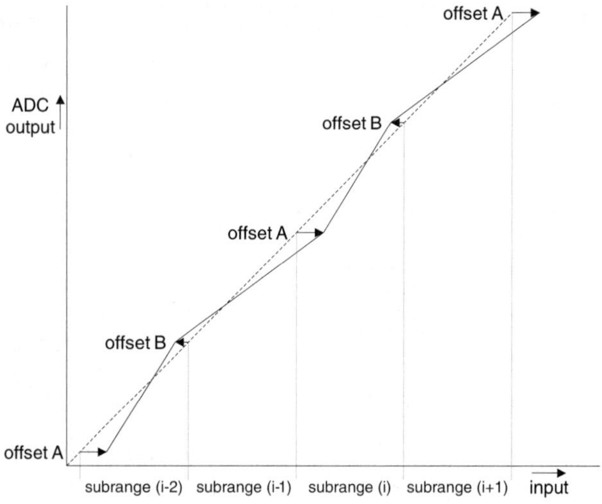

Figure 5.4: *The effect of residue amplifier offset with improved switching on the overall transfer curve (ignoring over-range)*

linearity of the A/D converter. Figure 5.4 shows that in the vicinity of a certain subrange transition the offset of only one amplifier (for example amplifier A) is relevant. Near the next subrange transition, the integral linearity of the A/D converter is mainly determined by the offset of the other amplifier (amplifier B). Since only 2 amplifiers are used to amplify the residue signals, some additional power and area can be spent to reduce the offset on these amplifiers. This is done by offset compensation, which is explained in the circuit design section.

Due to the offset compensation, the accuracy of the two-step A/D converter is determined mainly by the accuracy of the coarse reference ladders from which all signals are taken.

5.3 Circuit design

The design of the critical circuits is described in this section. As the A/D converter is designed for embedded use, the circuits are designed differentially to be less sensitive to any disturbance coming from the substrate or supplies. First the track-and-hold circuit is described, followed by the combination of coarse A/D converter, D/A converter and subtractor, the reference ladder requirements, the offset-compensated residue amplifiers and, finally, the fine A/D converter.

5.3.1 Track-and-hold circuit

The pseudo-differential track-and-hold circuit performs the sampling of the analog input signal at the input of the A/D converter. A single-ended channel is shown in figure 5.5.

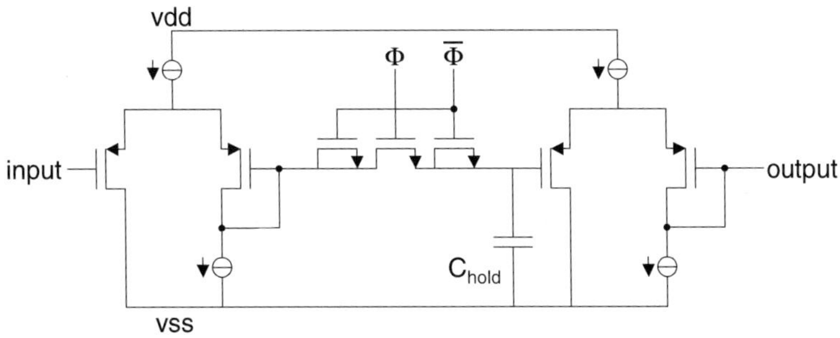

Figure 5.5: *Track-and-hold (single channel)*

The input signal is 0.8 V peak-to-peak single-ended, at a common mode level of 1.0 V. The low DC level requires the use of NMOS switches that have sufficiently low on-resistance. In order to prevent the overall SNR of the A/D converter from being affected by the thermal noise (equation 2.8), a sampling capacitor of 500 fF is used. The track-and-hold uses only one quarter of the total sampling period for tracking to maximize the settling time in the A/D converter. With a sample rate of 25 MSample/s, this equals 10 ns. The output signal of the track-and-hold has to track a Nyquist-rate input signal to an error of less than 0.25 LSB of 10 bits within 10 ns. Therefore 8.3 τ equals 10 ns. This means that the analog bandwidth of the combination of buffer, switch and sample capacitor has to be more than 135 MHz. Dummy transistors are added to compensate for the charge dump of the switch transistor [118].

5.3.2 Coarse A/D, D/A converter and subtractor

In order to generate the zero-crossings for the coarse A/D converter, the sampled signal from the track-and-hold is applied to two floating resistor ladders. This is shown in figure 5.6.

The comparators connected to this coarse ladder determine the coarse bits. The output signals of the comparators drive a digital XOR function, which selects the

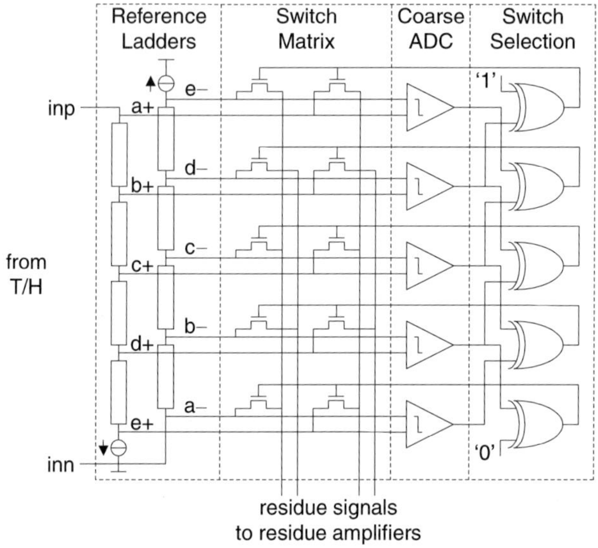

Figure 5.6: *Schematic of the coarse A/D converter, D/A converter and subtractor*

proper switches connected to the coarse reference ladders to generate the proper residue signals for dual-residue signal processing. These XORs and switches are also shown in figure 5.6. The switches consist of NMOS transistors. The generation of the dual-residue signals is explained in figure 4.5 and figure 4.6, without over-range.

5.3.3 Coarse ladder requirements

The coarse resistor ladder determines the integral linearity of the A/D converter because the signals used for the fine conversion (i.e. the residue signals) are derived from this ladder. Poly-silicon with a sheet resistance of 140 Ω is used as resistive material. The capacitive load of the comparators and the switches connected to the ladder together with the required speed determine the total resistance of the ladder. In poly resistors, the contact of metal-to-poly and the interface resistance from low-ohmic to high-ohmic poly is sensitive to process spread [119]. The connection of the ladder to the comparator inputs, as shown in figure 5.7, is made in such a way as to ensure that there is no current flowing through metal-to-poly contacts and low- to high-ohmic poly, and that their resistance does not affect the accuracy of the ladder.

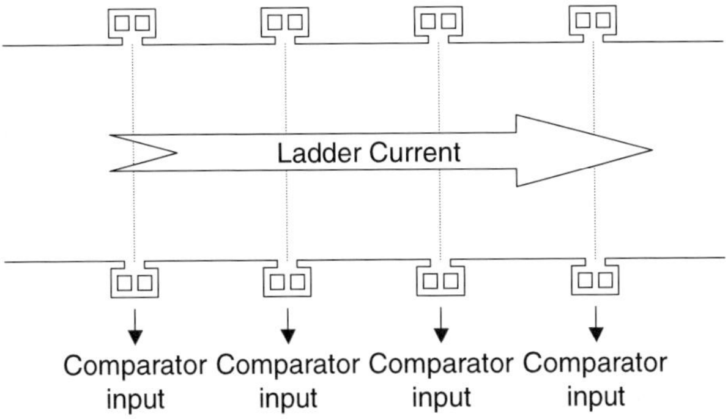

Figure 5.7: *One single-ended coarse and D/A converter reference resistor ladder without current through poly-to-metal contacts*

5.3.4 Offset compensated residue amplifier

Before the residue signals are applied to the fine A/D converter they are amplified. This is done by 2 differential amplifiers. Because the coarse A/D converter resolves 4 bits, the total input range is divided into 16 subranges. This allows a residue gain of 16 to generate the same amplitude at the output of the residue amplifier compared to the total input signal range. However, one residue amplifier remains connected to the reference ladder during 2 subranges (figure 5.6), which means that the gain factor has to be reduced to 8. In this design the gain factor is 10 because the distortion requirements after the amplifier are less stringent than at the input of the A/D converter. Since the offsets of the residue amplifiers make a major contribution to the linearity of the A/D converter, they have to be minimized. An online analog offset calibration [106] is implemented to reduce the capacitive load of the input stages of the residue amplifiers. Since the amplifiers cannot be used during calibration, the compensation is performed during the coarse quantization because then the amplifiers are not active in the signal path.

Both residue amplifier A and amplifier B consist of the amplifier shown in figure 5.8, which is based on a degenerated differential pair with gain boosting to increase the linearity [120]. When the switches connected to Φ_1 are closed (Φ_1 is high), the amplifier is in amplification mode and its input signal and offset are amplified. During calibration, the amplifier is disconnected from the switch unit and the input of the amplifier is short-circuited (Φ_1 is low). During this period only the offset appears amplified at the output. Switch Φ_2 is then closed and the amplifier enters negative feedback mode. This means that the amplifier forces its input signal, and also the output of the amplifier, to be equal to the offset divided

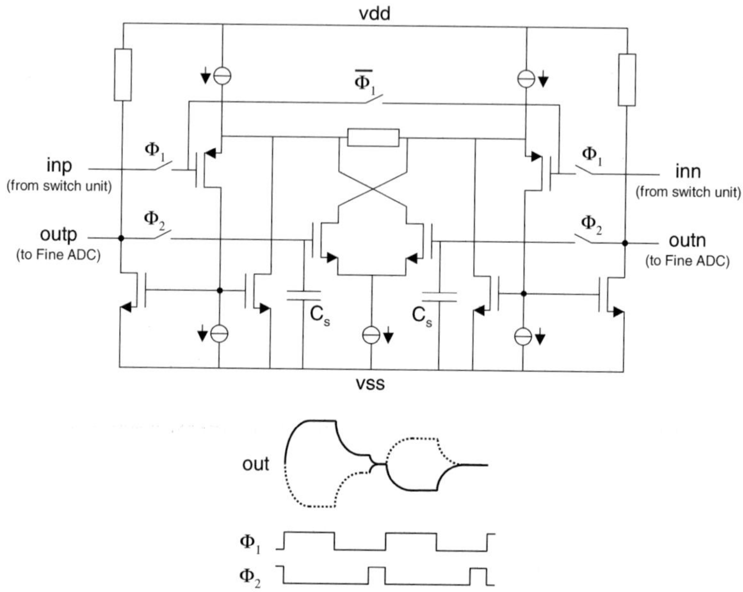

Figure 5.8: *One of the two residue amplifiers with offset calibration*

by the gain of the loop. The initial offset is reduced by the loop-gain of 10. This means that the area and therefore input capacitance of the input transistors can be reduced by a factor of 100 (equation 2.6), thereby reducing the capacitive load for the reference ladders. To have a stable feedback loop, the pole formed by the compensation amplifier has to be at a low frequency. The dominant pole of the feedback loop is determined by the storing capacitor C_s and the on-resistance of the switch Φ_2. The bandwidth of the offset compensation loop is small but because only static offsets have to be compensated this is not limiting.

5.3.5 Fine A/D converter

The conversion is completed in the fine A/D converter. OPAMP buffers apply the amplified residue signals to 2 resistor ladders to generate the zero-crossings by interpolation for the fine conversion. Until now only the circuits without over-range were considered. Increasing the number of over-range levels reduces the total capacitance of the coarse A/D converter, but increases the capacitance of the fine A/D converter. The optimum value is found when the fine A/D converter generates an additional bit, as is discussed in section 3.4. In order to increase the fine A/D converter by an additional bit, half an additional range is required at each side of the in- range part. The over-range zero-crossings of the fine converter are generated by placing resistors in the feedback-loop of the OPAMP buffers. Because this additional range is inside the feedback path of the OPAMP, as is shown in figure 5.9, this has a negative effect on the stability of this OPAMP. The OPAMP can be made stable with a large amount of over-range but this does involve excessive power. A better option when this effect is taken into account is to add 10 comparators over-range at each side in the fine A/D converter. The increase in allowed offset in the coarse A/D converter can be calculated using equation 5.2:

$$\frac{\sigma\left(V_{offset,coarse,\#ORC=10}\right)}{\sigma\left(V_{offset,coarse,\#ORC=0}\right)} = \frac{\frac{1}{4}+10}{\frac{1}{4}} = 41 \tag{5.3}$$

This results in a reduction in capacitive load in the coarse A/D converter of 41^2. The total fine A/D converter is shown in figure 5.9.

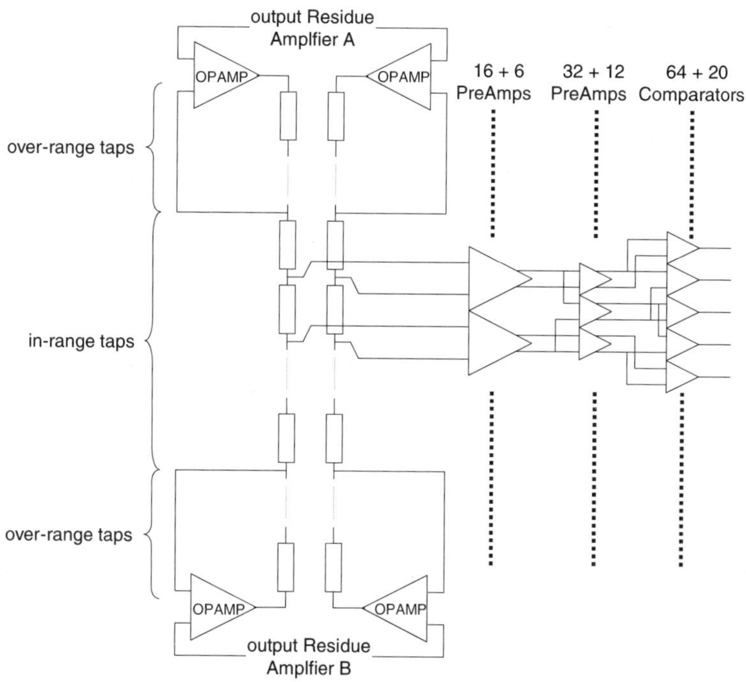

Figure 5.9: *Fine ADC with over-range generation and interpolation*

The 16 in-range and 6 over-range pre-amplifiers amplify the resistor ladder tap signals by a factor of 3. The outputs of these amplifiers are connected to 32 in-range and 12 over-range pre-amplifiers. Taking the outputs of adjacent pre-amplifiers generates the additional zero-crossings [15] (3.2.2). After these pre-amplifiers, 32 in-range and 12 over-range zero-crossings are generated. Again, by interpolation, 64 in-range and 20 over-range zero-crossings are detected by comparators to convert the analog zero-crossings to digital signals. The 64 comparators compare the in-range zero-crossings, while the extra 20 comparators compare the over-range zero-crossings of the fine A/D converter. Using this interpolation method, only a total of 22 pre-amplifiers are connected to the fine resistor ladder, limiting the capacitive load and thus increasing speed or reducing power. Due to the gain of the pre-amplifiers in front of the comparators, the effect of the comparator offset is also reduced, which has a positive effect on the DNL of the A/D converter.

5.3.6 Timing

The timing in the two-step A/D converter is determined by the required settling times of the various blocks. The timing is shown in figure 5.10.

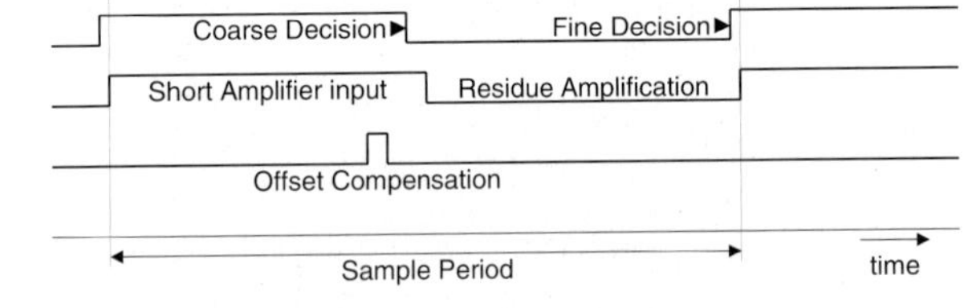

Figure 5.10: *Timing of the A/D converter*

During the track time, the analog input signal is applied to the A/D converter. The track-and-hold samples this signal and holds it for the remaining part of the sample period. After the coarse quantizer has settled, the coarse decision can take place. Using this result, the proper switches are set in the switch unit. This also needs some time to settle. Next, the residue amplifiers are switched on and amplify the residue signals. These signals are applied to the fine quantizer where, after some settling time, the final fine decision is taken. After this decision, the track-and-hold can track a new sample and the amplifier inputs are shorted. The amplifier offset-compensation takes place after the output of the amplifier has settled to zero. This cycle is repeated every 40 ns (25 MSample/s).

5.4 Experimental results

All the measurements in this book have been done using the test setup shown in figure 5.11.

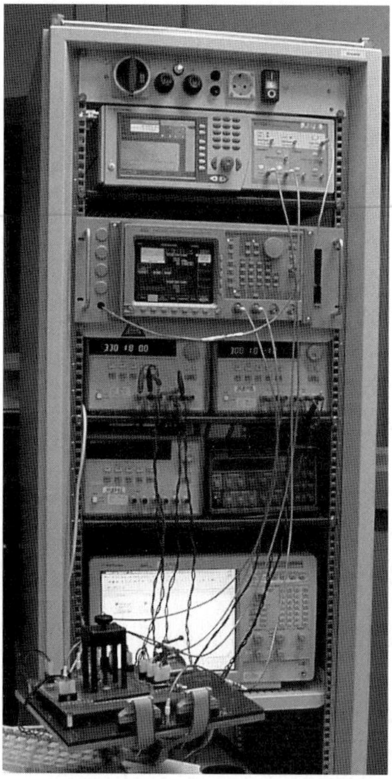

Figure 5.11: *Photograph of the test setup for characterizing the A/D converters designed in this book*

The supply sources are *HP3631A* and the reference current sources are *Keithley 224*. The clock reference for signal generation, A/D converter clock generation and data capturing are generated by the *Agilent 81134A* (formerly *Agilent 8133A*). The signal generator is an arbitrary waveform generator (*Tektronix AWG2021*).

The output data from the A/D converter is captured by a logic analyzer (*Agilent 1682AD*). All the equipment is set by a Labview program, which also does the signal analysis. A screenshot of the Labview program is shown in figure 5.12.

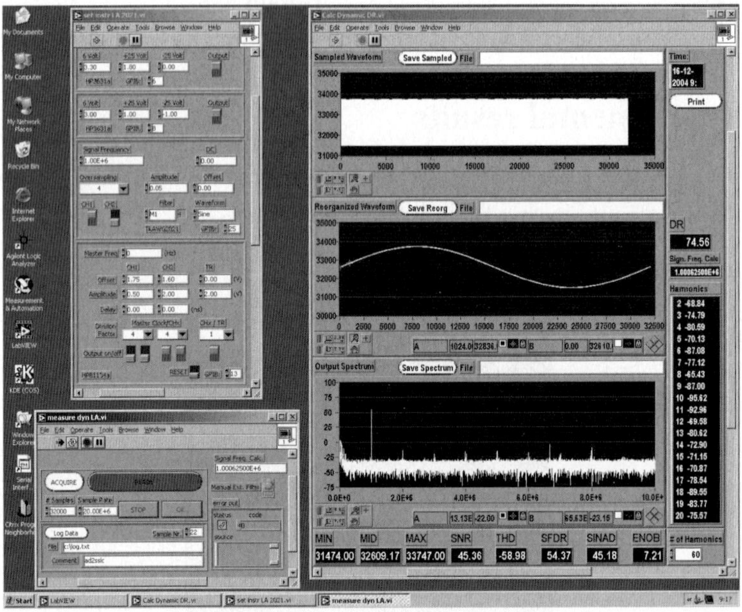

Figure 5.12: *Screenshot of the Labview program for equipment settings and signal analysis*

The screenshot of figure 5.12 is used for the 16-bit A/D converter from chapter 7. Each A/D converter has it own equipment configuration and Labview program. A separate analysis program exists for the dynamic and static measurements.

The two-step A/D converter has been implemented in a 0.35μm standard digital CMOS technology. No extra options, such as double-poly capacitors, have been used. The A/D converter operates with a supply voltage of 3.3 V and has a power consumption of 195 mW. The power consumption of each block separately is not available. The area of the A/D converter is 0.66 mm^2 including the track-and-hold and clock generation circuit. A die photograph is shown in figure 5.13.

The track-and-hold, reference ladders and coarse comparators are located at the top left-hand side. To the right side of these are the offset-compensated amplifiers and fine ladder buffers. The fine A/D converter with 64 + 20 comparators is situated at the bottom of the block. The digital decoder is located at the top right-hand side.

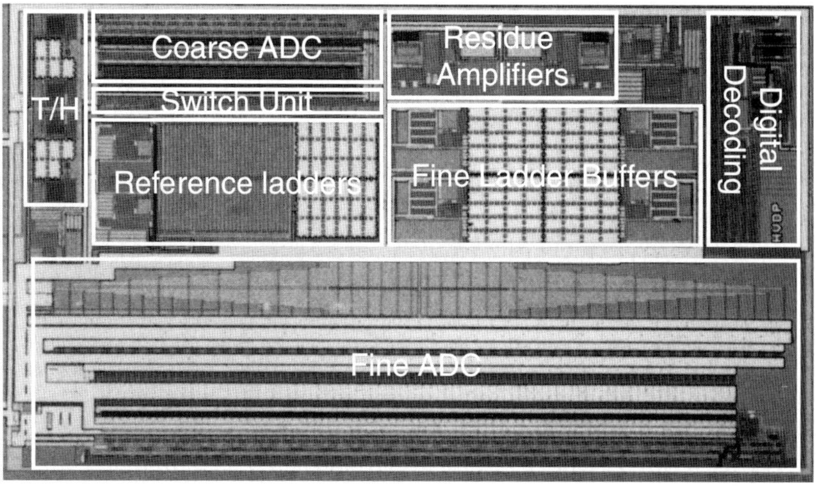

Figure 5.13: *Die photograph of the 10-bit A/D converter*

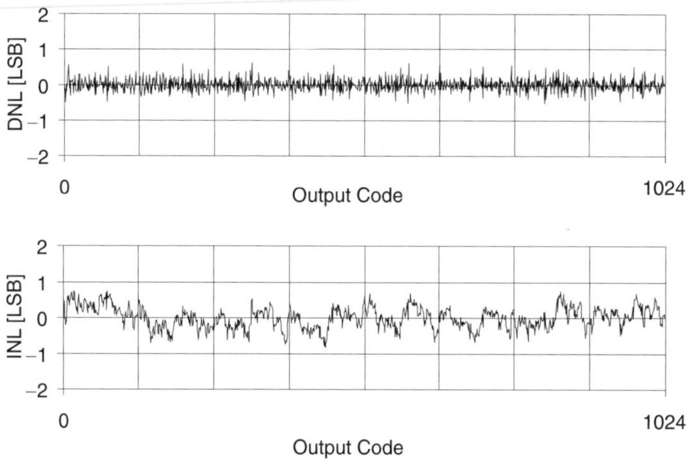

Figure 5.14: *Measured DNL and INL (1 MHz, 25 MSample/s)*

Figure 5.14 shows the DNL and INL measurement obtained at a sample rate of 25 MSample/s and an input signal frequency of 1 MHz. The DNL and INL are 0.7 LSB and 0.9 LSB respectively. Figure 5.15 shows the output spectrum of the A/D converter with a 1 MHz full-scale input signal, sampled at 25 MSample/s. This figure shows that, due to the offset compensation and the improved switching, the spurious components are below 72 dB. Figure 5.16 shows measurements

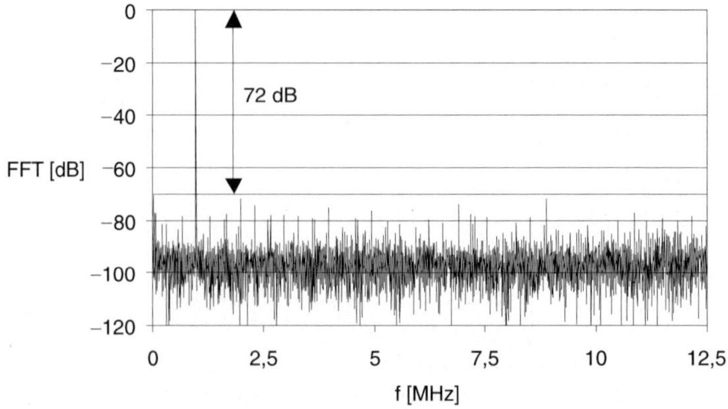

Figure 5.15: *Measured output spectrum (1 MHz, 25 MSample/s)*

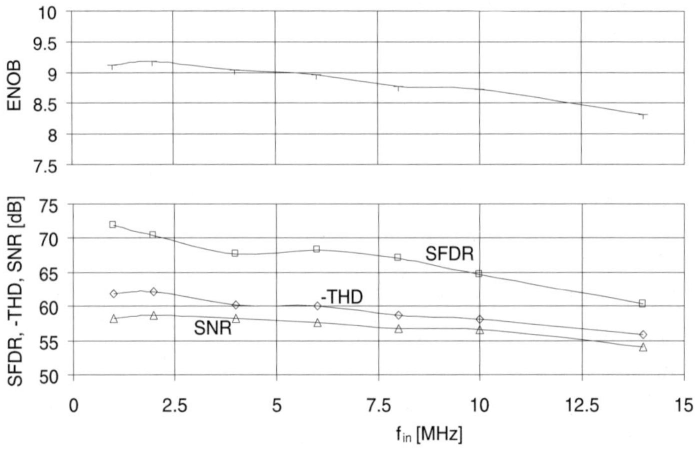

Figure 5.16: *ENOB, SFDR, -THD and SNR as a function of the input signal frequency (25 MSample/s)*

of SFDR, total harmonic distortion (THD) and SNR as a function of the input signal frequency with a sample rate of 25 MSample/s. The performance slightly degrades for higher signal frequencies. The ENOB is 9.0 at DC and more than 8.5 for input signals up to 12 MHz, resulting in an effective resolution bandwidth (ERWB) of 12 MHz. The *FoM* calculated with equation 3.35 is equal to 14,21 pJ/conv. When normalized to the input voltage amplitude equation 3.37, it equals: 36,39 pJV^2/conv.

Table 5.2: *Overview of the performance measured*

Technology	Single poly 0.35 μm CMOS
Resolution	10 bit
Supply	3.3 V
Sample rate	25 Msample/s
Input range	1.6 V_{pp}
ERWB	12 MHz
ENOB	9
SNR	58 dB
THD	−62 dB
SFDR	72 dB
INL	0.9 LSB
DNL	0.7 LSB
Power	195 mW
Area	0.66 mm^2

5.5 Discussion

No special care has been taken in creating an exact residue amplifier gain factor and the offsets of the residue amplifiers have been reduced by a factor of 10, which is sufficient to achieve a DNL and INL of 0.7 LSB and 0.9 LSB respectively. The DNL and INL before calibration is not available since in this realization the calibration cannot be switched off. The DNL is most probably limited to 0.7 LSB by the offsets of the comparators and pre-amps in the fine A/D converter. In order to improve this larger input devices would be required for the comparators, which would degrade the speed or increase the power. A small residual offset component can be distinguished in the INL measurements, as an 8 times repeating pattern, but the INL is still below 0.9 LSB. A larger offset compensation loop gain would be required to improve this. The SFDR is 72 dB for a 1 MHz signal. The SFDR and THD are a consequence of the INL curve.

The limitations of the proposed technique are identified as insufficient calibration loop gain and required calibration time. The solving of these limitations in the analog calibration circuits always results in a trade-off between speed and accuracy versus power. A fundamental problem in this technique is that the DC error in the amplifiers is corrected every sample clock cycle. This imposes unnecessary bandwidth requirements on the calibration loop and requires an additional fraction of each conversion cycle. In this implementation this additional time fraction results in a reduction of a factor of two in sampling speed. This can be seen in figure 5.10, where half of the sampling period is used for calibration of the residue amplifiers. This increases the achievable $FoM_{V_{pp}}$ by a factor of 2. Since in this experiment the A/D converter has not been optimized for power, the $FoM_{V_{pp}}$ is larger than expected from figure 3.33. The power could be reduced by optimizing in the second stage, for example, using a pipe-line 6-bit converter. The $FoM_{V_{pp}}$ can be further improved by reducing the power of the several building blocks in the converter and using more advanced technology.

If the measurement results are compared with the video requirements as defined in section 5.1, this A/D converter is shown to meet the specifications. The A/D converter achieves over 9 ENOB at DC and has a SFDR of over 72 dB at a sample rate of 25 MSample/s. The effective resolution bandwidth is close to the Nyquist frequency of 12.5 MHz. The A/D converter is fabricated in a 0.35 μm standard CMOS technology without options; it measures 0.66 mm^2 and consumes 195 mW at a supply voltage of 3.3 V.

5.6 Conclusions

In the experiment shown in this chapter the dual-residue signal processing in combination with analog offset calibration have been verified experimentally. Compared with the benchmark values in chapter 3, the power consumption is still too high, but the performance meets the requirements for CVBS video signal processing.

The limitation of the present calibration implementation is the result of the combination of A/D conversion and offset cancellation in one clock cycle. The next chapter will describe a realization which circumvents this trade-off.

Chapter 6

A 12-bit two-step ADC with mixed-signal chopping and calibration

6.1 Introduction

The previous chapter shows that the speed limitation of analog compensation becomes significant when considering figure 5.10. Due to the additional calibration time required, the maximum speed is limited to twice the minimum achievable settling time. The scope of the experiment described in this chapter is to verify the advantages of the dual-residue signal processing [3] from section 4.3.2 in combination with mixed-signal chopping and calibration (MSCC) from section 4.5 in the two-step architecture [26] (section 3.4). A GSM base-station application has been used as a test vehicle. In the early phase of GSM development, digitization took place on the demodulated baseband signal, see figure 6.1a. This position of the A/D converter in the signal chain relaxed the requirements on the A/D converter performance, however it imposed some stringent requirements on the system solution as a whole, e.g. channel selection had to be performed by setting the local oscillator. Moreover, this set-up required a full PCB for every channel. In a second phase, the A/D converter moves to the IF domain, thereby moving large parts of the filters into the digital domain. This A/D converter still serves one channel, however more bandwidth and dynamic range are required, figure 6.1b. Ultimately the A/D converter will have to convert all channels in the GSM band simultaneously. This will make it possible to move all channel-specific functionality into the digital domain, and also to adapt relatively easily to new standards that occur in the same band (e.g. the EDGE protocol). Since there can be large signals together with small signals, this requires a large A/D converter dynamic range. It is obvious that excellent spurious free dynamic range performance is required and this directly translates into a very high specification on the overall accuracy of the conversion function figure 6.1c.

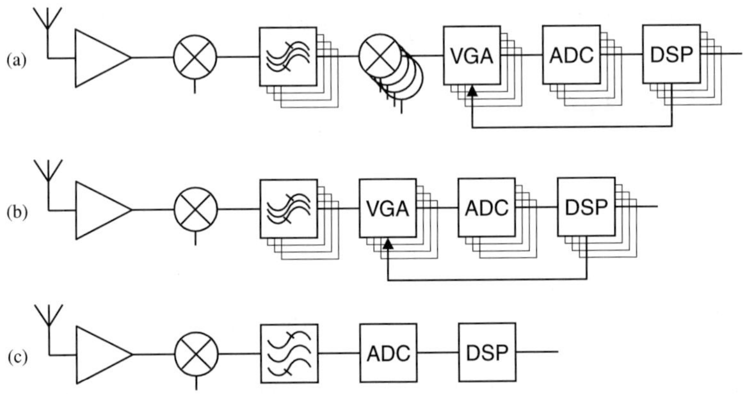

Figure 6.1: *GSM base-station receiver with: baseband A/D conversion (a), single-channel IF A/D conversion (b) and multi-channel IF A/D conversion (c)*

The trend in this application is to digitize a large number of channels with one A/D converter. The intermediate step, as shown in figure 6.1b where each channel in the IF-band is digitized with separate conversion chains, and a filter for each channel will serve as the system driver in this chapter. With respect to only baseband conversion, this removes the requirement for a separate mixer for each channel. The dynamic range of the A/D converter is relaxed because of the use of a channel filter and an AGC. The requirements of an A/D converter in such a system are shown in table 6.1.

Table 6.1 shows the general requirements for IF conversion for baseband signals. Cost requirements demand that the A/D conversion function is embedded in a digital CMOS chip. The GSM signal is down-converted to an IF frequency, resulting in a spectrum from 0 to 25 MHz in which the wanted channel has been pre-filtered. However, large components from interferers and strong adjacent signals can still be present. The 12-bit accuracy is sufficient together with the amount of channel filtering and the gain of the VGA. To be able to deal with the complete GSM band, the sample rate of the A/D converter has to be 50-60 MSample/s with

Table 6.1: *Requirements for IF conversion*

Technology	Digital CMOS without options
Resolution	12 bit
Supply voltage	Single supply acc. to architecture
Sample rate	>50 Msample/s
Effective bandwidth	25 MHz
SNR	>66 dB
SFDR	>75 dB
THD	>70 dB
Power dissipation	<400 mW
Area	<1 mm^2

an effective resolution bandwidth of 25 MHz. The SINAD of the GSM signal is only 9 dB, but in order to handle the large neighboring channels the SNR of the converter needs to be 66 dB. The spurious tones generated by large interfering unwanted channels can disturb the reception of a small wanted channel. The SFDR must therefore be below 75 dB. Distortion causes inter-modulation of large unwanted signals, the resulting products of which can fall in the wanted channel band. Embedding of such a system requires low power and low area. The target power dissipation is chosen as 300 mW.

This chapter describes the realization [58] of a two-step A/D converter using a mixed-signal chopping and calibration (MSCC) technique which reduces the residue amplifier offset without requiring additional calibration time. The performance improvements due to MSCC are achieved predominantly for sampling speed and INL and the corresponding linearity parameters such as THD and SFDR. Since the speed is improved by a factor of two with respect to the previous chapter, the sample rate achieved is over 50 MSample/s. The more advanced technology is also beneficial for speed improvement. The offset reduction of the MSCC is more effective than the analog calibration from the previous chapter, which allows the accuracy to be increased to 12 bits. The MSCC operates online, which means that the A/D converter continues to operate during calibration. The calibration works continuously; gradually changing offsets are tracked. The additional analog circuitry for calibration is limited and is not demanding. The digital properties of the CMOS technology are exploited by using digital offset extraction and storage. Large controlled time constants can be realized digitally, in a relatively small area.

Section 6.2 of this chapter describes the two-step architecture. It shows how the demands on the analog blocks in this design are minimized by architectural choices. Section 6.3 explains the mixed-signal and chopping and calibration algorithm, which is used to compensate the offset errors from the residue amplifiers. Some circuit designs are shown in Section 6.4. Section 6.5 and section 6.6 show the experimental results and the discussion respectively. Finally, the conclusions are presented in section 6.7.

6.2 Two-step architecture

A detailed diagram of the two-step A/D converter presented is shown in figure 6.2.

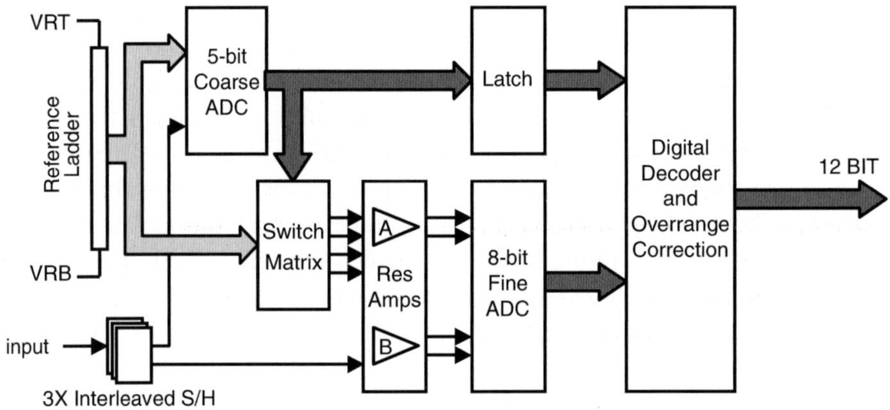

Figure 6.2: *Block diagram of the two-step A/D converter*

The differential input signal is sampled with three interleaved sample-and-hold circuits. The acquired analog signal is applied to the coarse A/D converter, which compares the differential input signal with a static reference ladder [121]. Comparison of the signal to a static reference ladder increases the allowable input signal swing by a factor of 2, compared with when the differential input signal is applied to 2 floating ladders, as described in chapter 5. The result from the coarse quantization is stored in a latch, and is also applied to a switch matrix. According to the coarse quantization, this switch matrix selects 4 reference signals from the same ladder as used for the coarse quantization. This is explained in section 6.2.3. These selected reference signals are combined with the held input signal in 2 residue amplifiers. The residue amplifiers perform the residue generation and provide a gain of 8 before the residue signals are applied to the 8-bit fine

A/D converter, which is explained in section 6.2.4. The accuracy requirements of the fine A/D converter are relaxed by providing gain to the residue signals. Since there are no error-contributing components up to this point in the signal path, except for the sample-and-hold and the static ladder, the offset generated in these amplifiers is the main accuracy-limiting component in the A/D converter.

The optimum distribution of the coarse and fine bits found in section 3.4 suggests a 6-bit coarse A/D converter, although an increase in the number of coarse bits would not only increase the load of the coarse A/D converter by a factor of 8, but would also increase the number of switches for the fine reference selection. This would cause the load of the reference ladder to increase. Decreasing the number of coarse bits increases the accuracy requirements of the fine A/D converter, since it has to generate 9 bits instead of 8. The distribution for this A/D converter is set at a 5-bit coarse and an 8-bit fine A/D converter.

Since the fine A/D converter has half a subrange over-range available on both sides of a subrange, the accuracy of the coarse A/D converter only has to be at half an LSB of its resolution, which equals 6-bit accuracy. The 8 bits of the fine A/D converter can be generated with sufficient accuracy without using compensation by using a folding and interpolating A/D converter. In the following paragraphs building blocks are discussed in more detail.

6.2.1 Interleaved sample-and-hold

The use of interleaved sample-and-holds together with a static reference ladder enables pipe-lining in the A/D converter. The total sample-and-hold consists of 3 identical interleaved [122] sample-and-hold circuits, as shown in figure 6.3.

The sample-and-holds all have the same input signals, while the clock signals are interleaved in time. The clock signals are 'one-high out of three'. This means that each sample-and-hold samples during one clock period, and holds during two clock periods. Each step in the A/D converter has a full sample period settling time available, while settling to full 12-bit accuracy is only necessary at the input of the residue amplifiers. This interleaving timing scheme enables high-speed sampling, since three actions (sampling, coarse and fine decision) can take place simultaneously on different analog samples. This interleaving is also shown in figure 6.3. If sample-and-hold 1 (S/H1) is tracking the analog input signal (n), then S/H3 is carrying out the coarse quantization of the previous sample ($n-1$), and S/H2 is performing the final quantization of two samples earlier ($n-2$). The

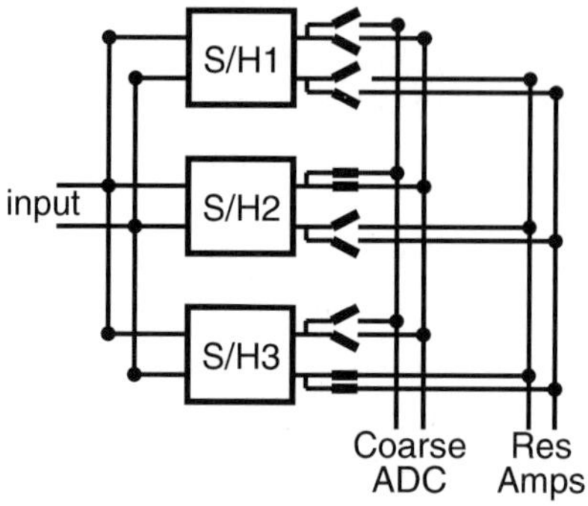

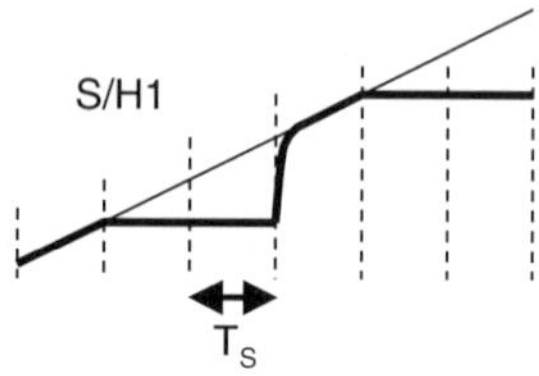

	Action		
S/H1	Sample (n)	Coarse (n)	Fine (n)
S/H2	Fine (n–2)	Sample (n+1)	Coarse (n+1)
S/H3	Coarse (n–1)	Fine (n–1)	Sample (n+2)

Figure 6.3: *Interleaved sample-and-hold*

signal is sampled on a capacitor and remains there for both the coarse and fine conversion steps. There is no charge transfer or re-sampling during coarse and fine quantization and amplifying. This property is good for noise considerations, as wide-band noise sampling only occurs in the S/H and in the fine quantization. It is obvious that the matching of the three interleaved sample-and-holds is very important [123]. This is discussed in section 6.5 and section 6.6.

6.2.2 Coarse A/D converter

The differential input signals are compared with a static reference ladder to obtain the coarse quantization, as shown in figure 6.2. In order to compare 2 signals with a static reference ladder, pre-amplifiers with 4 inputs are used [121]. 2 inputs are connected to the output of the sample-and-hold and 2 inputs are connected to the reference ladder, as is shown in figure 6.4.

All coarse A/D converter zero-crossings are generated by using different references for each pre-amp. The accuracy requirements of the 5-bit coarse A/D converter is limited to only 6 bits, as the fine A/D converter is able to correct errors of up to half a subrange (section 3.4).

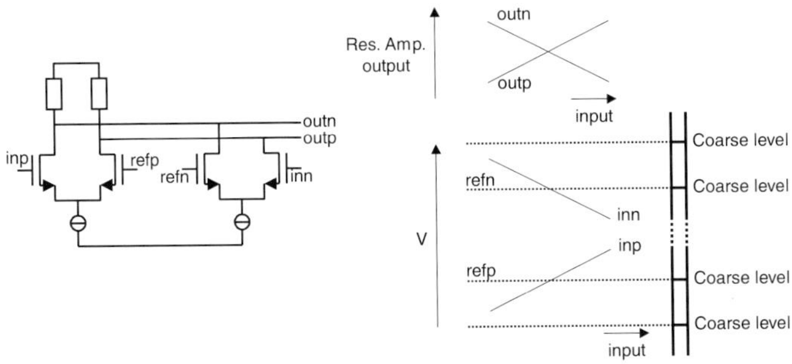

Figure 6.4: *Pre-amplifier with 4 inputs to generate zero-crossings at different references*

6.2.3 Switching and residue signal generation

The coarse decision generates a thermometer code and an XOR block is used to generate a 'one high' code from this thermometer code. The input range of the A/D converter is divided into a number of sub-ranges (SUB). The amount of sub-ranges is equal to the number of coarse levels. The 'one high' code selects the proper sub-range in which the input signal is located. To be able to generate the proper (dual) residue signals for the fine quantization, the switch unit selects 4 references from the static reference ladder according to the 'one high' code. The 2 closest references to the differential sample-and-hold output signal and the 2 second closest references are selected. This is illustrated in figure 6.5a for subrange $SUB(n)$ and subrange $SUB(n+1)$.

In order to be able to use the improved switching approach as described in section 4.3.2 [91], *refB* and *nrefB* remain connected to the reference ladder when switching from subrange $SUB(n)$ to $SUB(n+1)$. The selected references *refA* and *nrefA* together with the differential output signal of the sample-and-hold are applied to residue amplifier A, while *refB* and *nrefB* together with the differential output signal of the sample-and-hold are applied to residue amplifier B. These amplifiers (which will be explained in the next section) combine their input signal according to the following equation:

$$\begin{aligned} resA &= inp - refA \\ nresA &= inn - nrefA \\ nresB &= inp - refB \\ resB &= inn - nrefB \end{aligned} \tag{6.1}$$

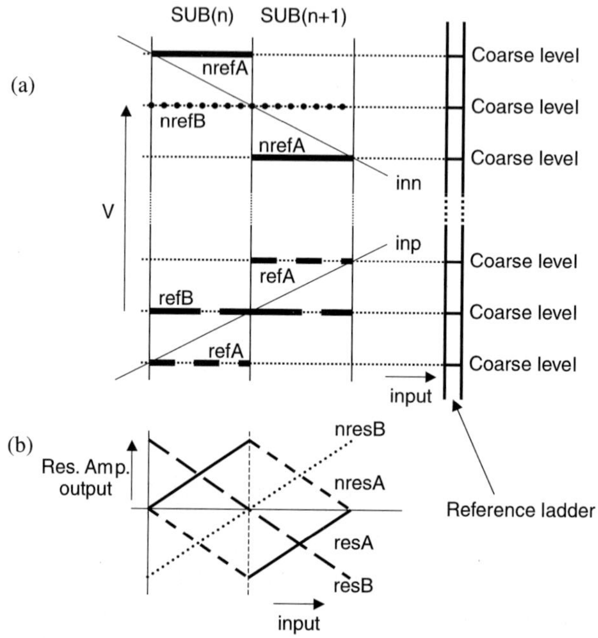

Figure 6.5: *Signal switching without over-range (a) and the residue amplifier outputs without over-range (b)*

In equation 6.1, *resA* and *nresA* are the differential output signals of residue amplifier *A*, while *resB* and *nresB* are the differential output signals of residue amplifier *B*. The result of this is shown in figure 6.5b. Both pairs of residue signals are applied, via buffers, to 2 floating resistor ladders of the fine A/D converter which generate the required fine quantization levels, as will be explained in section 6.4. The fine A/D converter range of *SUB*(*n*) begins at the zero-crossing of the signals *resA* and *nresA* and ends at the zero-crossing of the signals *resB* and *nresB*. The subrange transition from *SUB*(*n*) to *SUB*(*n* + *1*), as shown in figure 6.5, is determined by the coarse quantization and in the case of figure 6.5 is ideal. However, when the coarse quantization is not accurate, the subrange transition is at another position. This is illustrated in figure 6.6a.

Because the fine A/D converter range is from the zero-crossings of *resA* and *nresA* to the zero-crossing of *resB* and *nresB*, there are no quantization levels at the right of the zero-crossing of *resB* and *nresB*, indicated by the circle in figure 6.6b, causing the fine A/D converter to clip. After the subrange transition from *SUB*(*n*) to *SUB*(*n* + *1*), the fine A/D converter range does not start at the zero-crossing

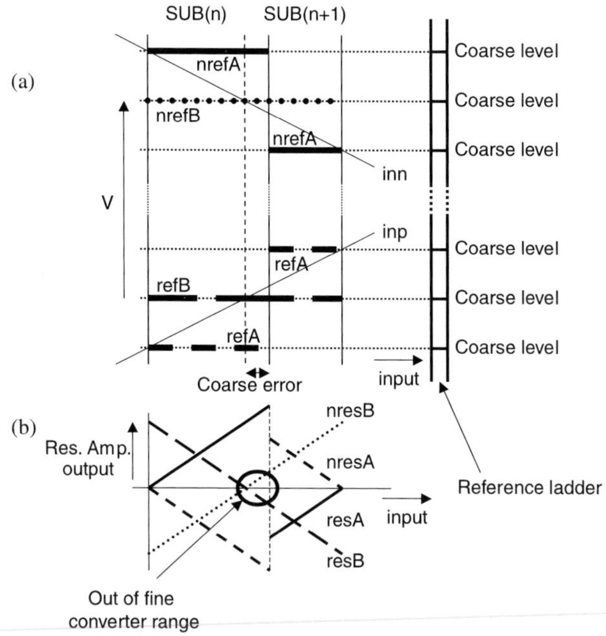

Figure 6.6: *Signal switching with coarse A/D converter error, without over-range (a) and the corresponding residue amplifier outputs (b)*

of *resB* and *nresB*, but has skipped part of the fine A/D converter range. The quantization levels left from the subrange transition are therefore not used. This causes missing codes in the transfer curve of the total A/D converter similar to figure 3.18 [29]. To circumvent the occurrence of missing codes, over-range is applied in the fine A/D converter (section 3.4). The fine A/D converter does not use the same references as are used by the coarse decision to generate this over-range (as was shown in figure 6.5), but connects to other reference taps in order to double the fine A/D converter range. This is shown in figure 6.7a.

In figure 6.7b the corresponding output signals of the residue amplifiers show that in this case the fine A/D converter input-referred range is from A to B. This creates an overlap in the subrange transition (see also figure 3.19) from C to B, which means that coarse errors can be corrected as long as they are between C and B. This coarse error correction (figure 3.20) is performed by the over-range correction of the digital decoder shown in figure 6.2.

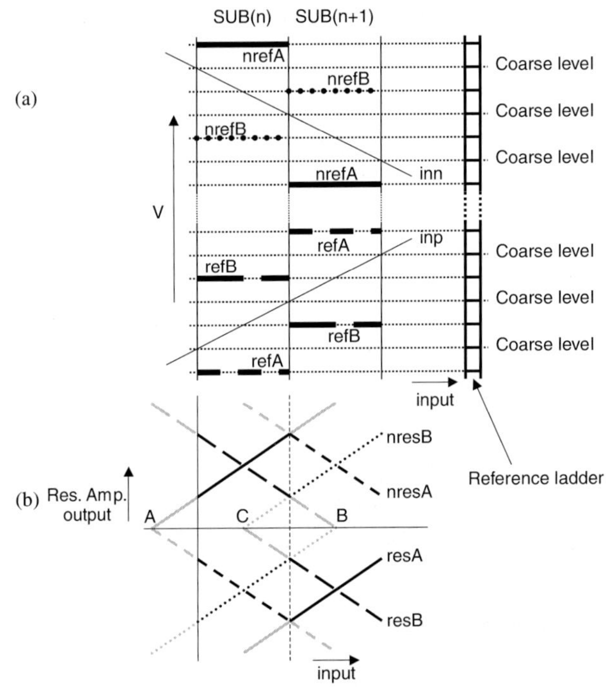

Figure 6.7: *Signal switching with over-range (a) and the corresponding residue amplifier outputs with over-range (b)*

Generation of the over-range part of the fine A/D converter by choosing the proper references overcomes the limitation described in chapter 5 of generating the over-range by inserting resistors in the feedback part of the fine resistor ladder buffers.

6.2.4 Residue amplifiers

The residue amplifiers are shown in figure 6.8.

The residue amplifiers have two functions. They act as subtractors to combine the output signals from the sample-and-hold and the selected reference signals to generate the residue signals which are applied to the fine A/D converter. The subtraction is done by two differential pairs similar to those in figure 6.4, where *inp* and *inn* are the output signals of the sample-and-hold and *refp* and *refn* are the selected references *refA* and *nrefA* for residue amplifier *A* and *refB* and *nrefB* for residue amplifier *B*, as described in the previous section. Their second function is to amplify the residue signals. This reduces the accuracy (noise and matching) requirements of all circuits after this amplifier by the gain factor. Because of the

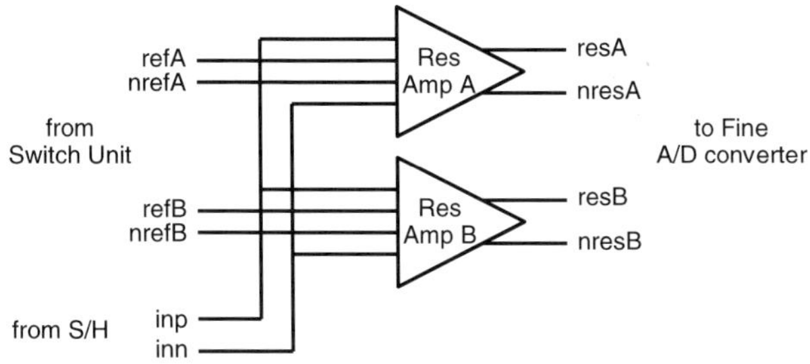

Figure 6.8: *The two dual-residue amplifiers*

use of dual-residue signal switching (section 4.3.2), the absolute gain requirement of the residue amplifiers is not important because this scales both the amplified residue signals and fine A/D converter range. The matching between both residue amplifiers is more important, but this is dependent on poly-resistor matching, as will be explained in the next section. The offsets generated in the residue amplifier mainly determine the linearity of the A/D converter. The next section describes a mixed-signal chopping and calibration algorithm to calibrate this offset.

6.3 Mixed-signal chopping and calibration

The dominant error-contributing components in the signal path before gain is applied are the sample-and-hold, the reference ladder, the switches in the switch unit, and the offset on the residue amplifiers. Sufficient power is expended in the sample-and-hold to meet the noise and linearity requirements. Matching of the reference ladder resistors is sufficient for 12 bits. The switches are just simple NMOS switches, designed to have a sufficiently low on-resistance. This provides sufficient bandwidth for 12-bit settling of the reference signals on the residue amplifiers. The offset on the residue amplifiers is now the only accuracy-limiting component. Reduction of the offset to better than 12-bit accuracy requires large devices, which limits the speed of the converter. Mixed-signal chopping and calibration (section 4.5) is applied to the residue amplifiers to allow the use of small device sizes and maintain speed and offset accuracy in the residue amplifiers. This calibration is performed on-line, which means that no additional calibration time is required and the converter is quantizing the input signal, while also carrying out the offset calibration. The digital processing capability of the CMOS technology is used to extract the offset from the A/D converter output. Digital signal processing allows filters to be created with large time constants. Since offset is an analog property, a calibration D/A converter is used to compensate for the offset on the residue amplifiers.

6.3.1 Residue amplifier offset

Due to the improved switching (section 4.3.2) [91], offset on both amplifiers affects the transfer curve, as shown in figure 6.9. The offset of only one of both amplifiers is shown at a subrange transition. This gives a deterministic, repeated pattern in the INL curve of the A/D converter, and the offsets can therefore be measured in the digital domain by observing the regular digital signal at the output of the A/D converter.

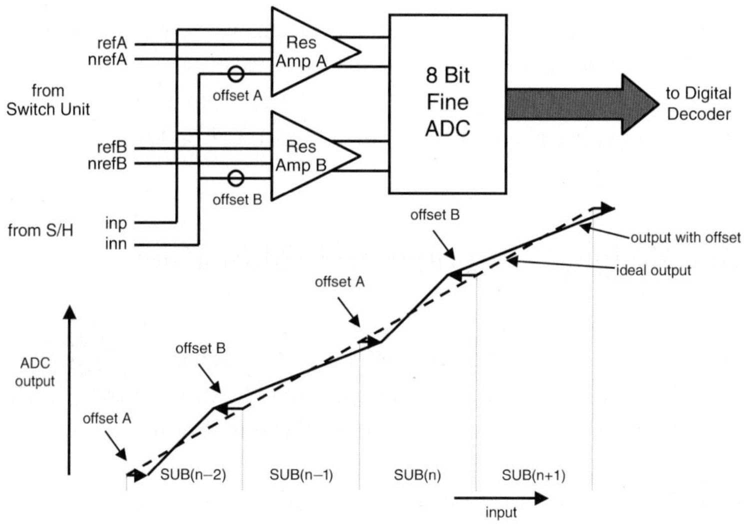

Figure 6.9: *Transfer curve of A/D converter with residue amplifier offset*

6.3.2 Chopping

To distinguish the offset of the residue amplifiers from the input signal of the A/D converter, chopping is applied to the inputs of both amplifiers (section 4.5). The required bandwidth in the amplifiers is not increased due to chopping because the chopping takes place at the same time as the subrange selection takes place. This is explained in section 4.5.

6.3.3 Digital extraction

Because in the case of two residue amplifiers the offset of both amplifiers is determined randomly, they can have any value. However, they can always be split up into two components: a common component, which is an equal value for both amplifiers, and a differential component, which is also equal but has the opposite sign (figure 4.24). A common offset component will only give a DC shift in the transfer of the A/D converter, and therefore seems harmless at first sight. However, such an error on the amplifiers will reduce the over-range capability in the fine A/D converter, which results in a smaller allowable offset in the coarse A/D converter. To extract both common and differential offset components the chopping method, as described in section 4.5.5, is applied. The total compensation loop is shown in figure 6.10.

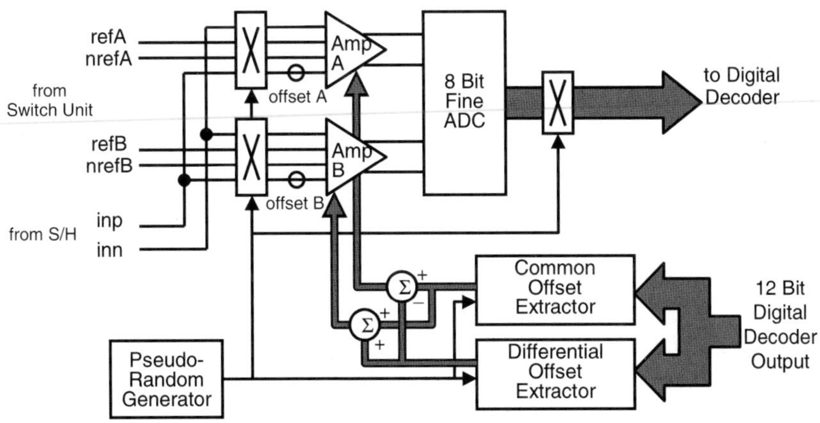

Figure 6.10: *Complete compensation loop with choppers, residue amplifiers, fine A/D converter and digital processing*

Both residue amplifiers have a chopper at the input. The signal is quantized in the 8-bit fine A/D converter after amplification by the residue amplifiers. In the digital domain the data is chopped back to retrieve the original input signal and is applied to the decoder to generate the 12-bit output of the A/D converter. This 12-bit output is applied to a common and a differential offset extractor. In this design not only the fine A/D converter output (as described in section 4.5), but the total 12-bit A/D converter output is applied to the digital offset extractor. This is not principally different. However, using only the fine A/D converter results in smaller required integration cycle lengths since the amplitude of the signal content is smaller (section 4.13). This signal is processed in order to retrieve the respective offsets at DC and by integration of the resulting signal the sign of both common and differential offset are extracted (section 4.5.5). The resulting sign is used to change the values of the digital signals which are applied to the offset compensation D/A converter located inside the residue amplifiers. The different effects of the common and the differential offset extractors can be seen at the ladders. A change caused by the common offset extractor gives both D/A converter values a step of the same sign, while a change caused by the differential offset extractor gives a step of the opposite sign. These D/A converters close the compensation loop, thereby removing the offset in the residue amplifiers.

The calibration algorithm presented (section 4.5) does not need a dedicated test signal and does not require a part of the conversion time. It works continuously and with every signal applied to the A/D converter. The major advantage of this mixed-signal chopping and calibration algorithm is that it extracts the errors from critical analog components in the digital domain, and compensates these errors in the analog domain.

6.4 Circuit design

6.4.1 Interleaved sample-and-hold

The sample-and-hold used in this design consists of three identical sample-and-hold circuits. These sample-and-hold circuits are interleaved in time [122], as described earlier. The sample-and-hold is shown in figure 6.11 [56, 33].

In the sample mode the single-stage folded cascode OPAMP is switched as a follower. This means that the DC value of the ladder is present on one side of the sample capacitor, in addition to the offset of the OPAMP. When the sample-and-hold switches to 'hold' mode, the sampling capacitor, which now contains the signal value and the offset of the OPAMP, is connected across the OPAMP. The offset is then applied to the capacitor with the reverse sign, thereby compensating for the sampled offset. The sampled value is seen at the output, with sufficiently low offset for 12-bit accuracy. The offset difference between the three sample-and-holds is very critical for high performance as well as the dynamic matching [123]. The latter is discussed in section 6.5 and 6.6.

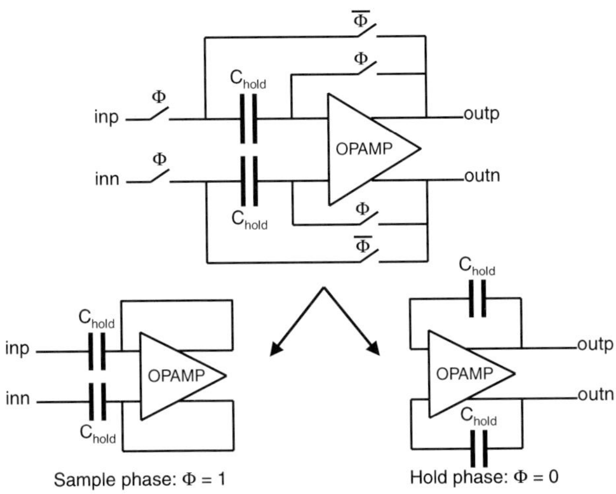

Figure 6.11: *Sample-and-hold circuit, split up into the sample phase and the hold phase, the OPAMP offset is canceled at the output*

6.4.2 Coarse A/D converter

The coarse A/D converter has to compare the two differential signals with the static reference ladder. 32 comparators generate 5 bits in the coarse A/D converter. Pre-amplifiers with interpolation (section 3.2.2) [15] are used because a lot of power would be required to drive the load of 32 comparators which have an accuracy of 6 bits at 54 MSample/s. Part of the coarse A/D converter is shown in figure 6.12.

There are 9 pre-amplifiers at the input, each with 4 inputs, two of which are the reference signals to which the input signal is compared. When the input signals equal their respective references, the output crosses zero. Interpolation is applied behind this first pre-amplifier stage; the additional zero-crossings are generated by combining the output signals from adjacent pre-amplifiers (section 3.2.2) [15]. Connected to these 9 four-input pre-amplifiers are 17 two-input pre-amplifiers, as shown in figure 6.12. Interpolation is applied again, and these amplifiers drive 32 comparators. The gain in the pre-amplifiers reduces the required accuracy, and thereby reduces the power consumption of the comparators. The fine A/D converter can correct for half a subrange error in the coarse A/D converter. This equals 28 mV, since the input range of the A/D converter is 1.8 V peak-to-peak.

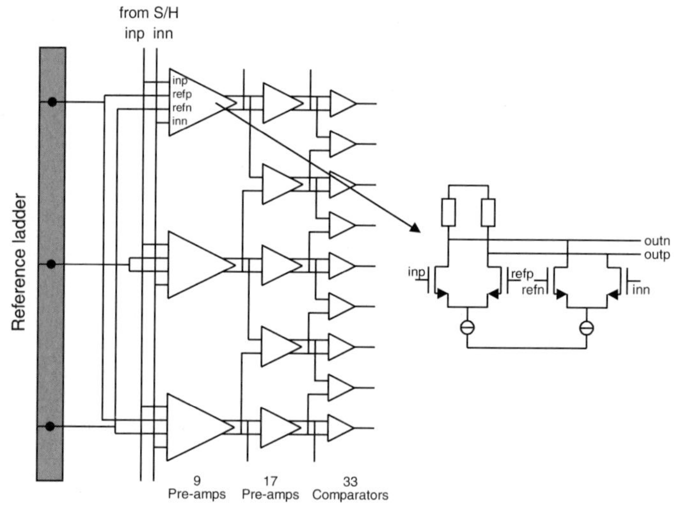

Figure 6.12: *Coarse A/D converter*

6.4.3 Residue amplifier with offset compensating current D/A converter

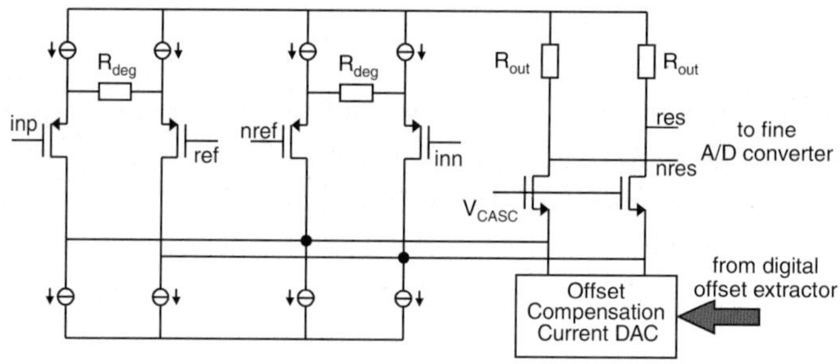

Figure 6.13: *Residue amplifier with compensation current D/A converter*

Figure 6.13 shows one of the residue amplifiers from figure 6.8. This amplifier has 4 inputs, which are needed to perform the subtraction of the input signal with the selected references from the reference ladder, to generate the residue signals. The input transistors are designed for a sufficiently low thermal noise contribution to the signal path. The degeneration resistors R_{deg} are added for sufficient linearity. The offset of this amplifier mainly determines the accuracy of the A/D converter. This offset is therefore calibrated by a current D/A converter. This D/A converter drives a current through the output resistors of the amplifier, such that the remaining offset of the amplifier is sufficiently low. The digital offset extraction block determines the digital code applied to the current D/A converter. Linearity of the D/A converter is not an issue since it is part of a feedback loop. The DNL of the D/A converter determines the resolution of the offset compensation. A 9-bit D/A converter was sufficient to compensate the largest possible offset to $\frac{LSB}{10}$ of the overall A/D converter. The absolute value of the gain of the amplifiers is not important, as is explained in section 4.3.2. The matching of the gain is determined by the matching of poly resistors and is sufficient for 12-bit accuracy.

6.4.4 Folding-and-interpolating fine A/D converter

The 4 residue signals from the residue amplifiers are connected to 4 OPAMPs switched as buffers, which apply their output signals to both the top and bottom of 2 moving ladders, as shown in figure 6.14.

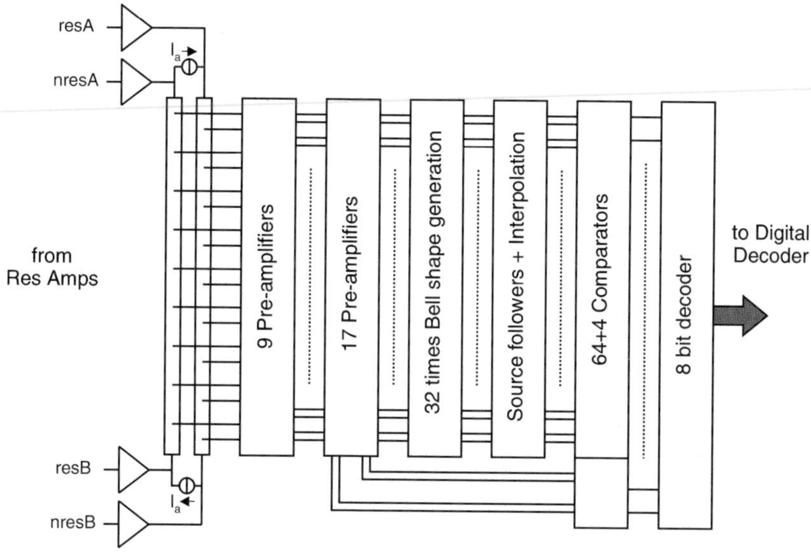

Figure 6.14: *Folding and interpolating fine A/D converter*

The direction of the current flowing in these ladders is determined by the selected subrange and by the chop state. Two buffers therefore have to sink the ladder current, and 2 buffers have to source the ladder current, depending on the subrange and chop state. The current is reversed in a subrange transition or a chop state change. This can give rise to a jump in the output signal of the buffers due to the large change in current. This effect can be reduced by introducing a large quiescent current in the output stage of the buffers, but this requires high power dissipation. The solution used here is to provide this switching current by using additional current sources, I_a shown in figure 6.14. They are connected at the top and bottom of both ladders to sink or source the ladder current. Now the buffers only have to deliver the error current, which is much less than the DC current through the ladders. The fine A/D converter shown in figure 6.14 has to provide 8 bits because a flash A/D converter would require 256 comparators (section 3.2), which would give too much load to the fine ladders. A folding and interpolating A/D converter is therefore used (section 3.3). The signal is folded 4 times, requiring 64 comparators. Most folding A/D converters use 8 or 16 times folding (section 3.3) [16, 19, 23], however 4 times folding was used since a 6-bit full-flash fine A/D converter from a previous design was expanded to 8 bits by only adding this 4 times folding block. Pre-amplifiers with interpolation [15] are used to generate the zero-crossings, as described in section 3.2.2, and used in the coarse A/D converter. Nine pre-amplifiers are connected to the ladder, followed by 17 pre-amplifiers. Connected to these pre-amplifiers are 33 folding amplifiers, which each generate a bell-shaped signal [21] by coupled differential pairs, as shown in figure 6.15.

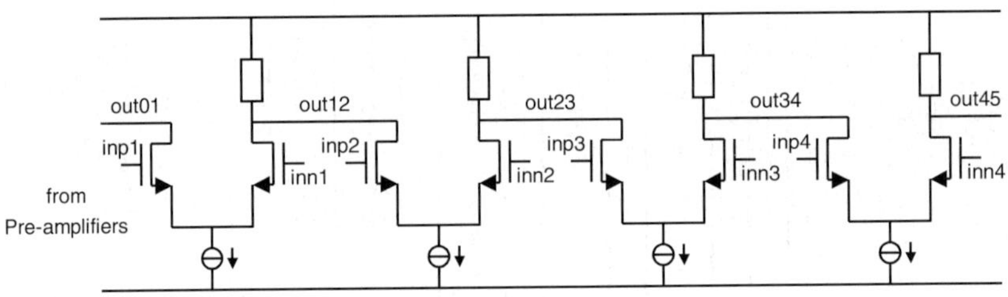

Figure 6.15: *Bell-shaped signal generation by coupled differential pairs*

These bell-shaped signals are combined with source followers [124], the sources of which are connected together. This is illustrated in figure 6.16 by the signal *out12* + *out34*, composed from combining *out12* and *out34*.

This generates 8 signals with 4 times folding. Then 64 zero-crossings are gengerated by resistive interpolation. Sixty-four comparators digitize these 64 zero-crossings. Four additional comparators generate the 'coarse' bits of the fine A/D converter.

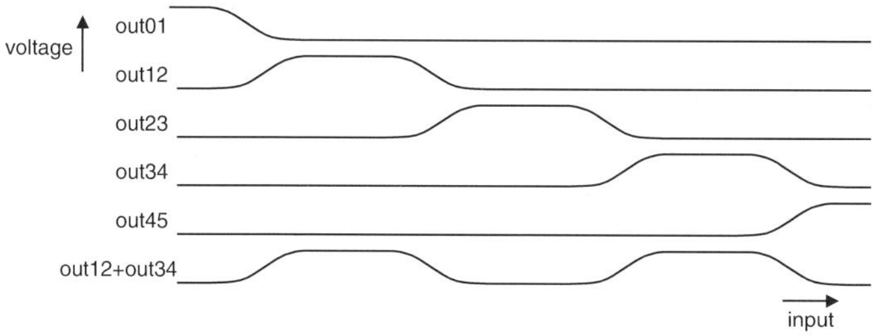

Figure 6.16: *Generation of the folding signal from combination of the bell-shaped signals*

6.5 Experimental results

The A/D converter has been fabricated in a standard 0.25 μm CMOS technology with one layer of poly silicon and five layers of metal. No additional options were applied (neither controlled capacitors nor thick gate oxide transistors). The supply voltage is 2.5 V. The area of the A/D converter is 1 mm^2. The die photograph is shown in figure 6.17.

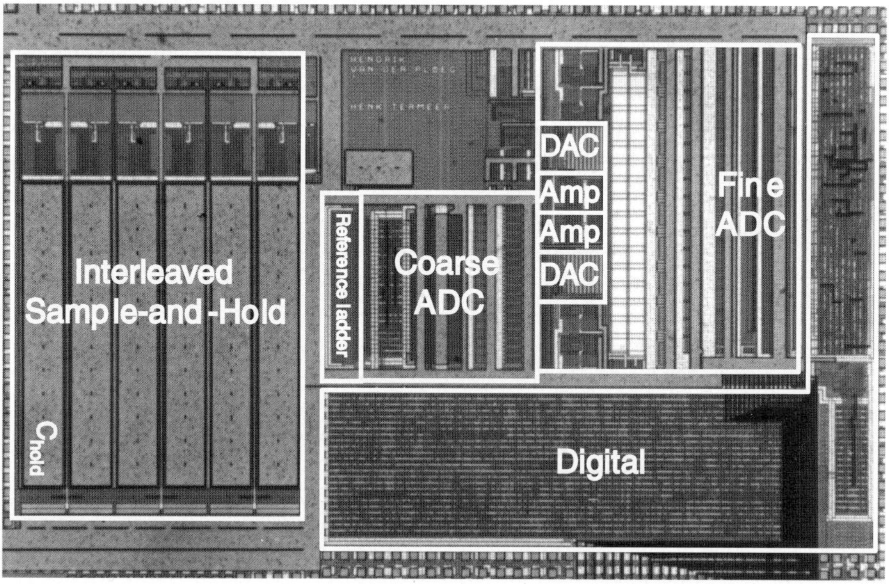

Figure 6.17: *Die photograph of the 12-bit two-step A/D converter*

The three interleaved track-and-hold blocks with their sample capacitors are on the left-hand side. The reference ladder and the coarse A/D converter are shown in the middle, in addition to the two residue amplifiers with the respective compensation D/A converters. The 8-bit folding and interpolating A/D converter with the buffers and ladders is on the right-hand side. The digital offset extraction block can be seen at the bottom.

The power consumption of the A/D converter is 295 mW (without output buffers). The power is divided over the different blocks of the A/D converter, as shown in figure 6.18.

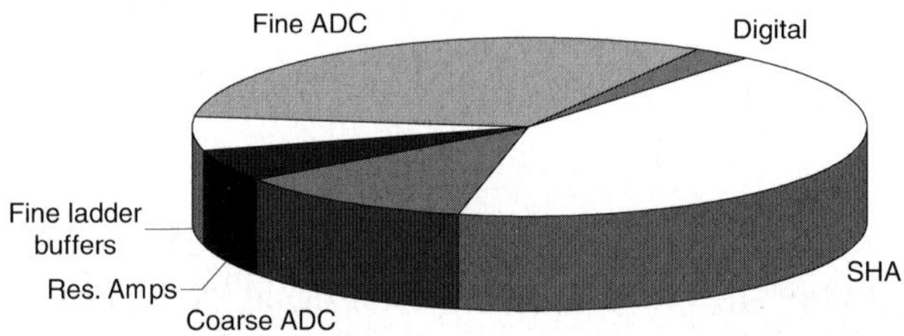

Figure 6.18: *The relative power consumption of each block of the A/D converter*

The major power-consuming blocks are the SHA and the fine A/D converter. As can be seen in figure 6.18, the power contribution of the residue amplifiers to the total power consumption is limited because of the offset compensation. The power for the digital circuits is only a small part of the total A/D converter power consumption and includes the digital decoder with over-range correction and the MSCC.

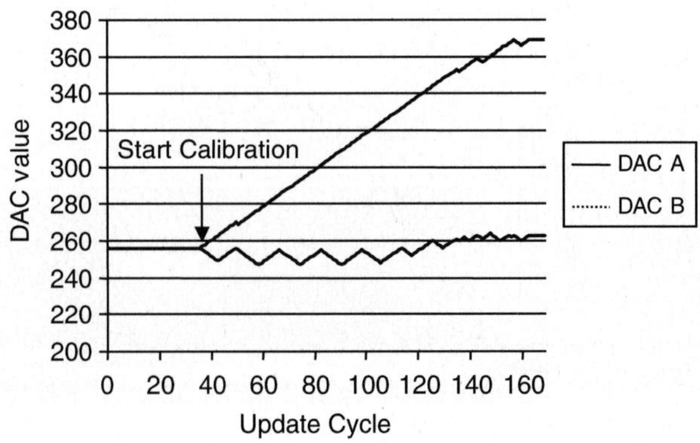

Figure 6.19: *Calibration of the residue amplifiers*

Figure 6.19 shows the calibration from start-up. Both D/A converter values are in the mid-range at the starting point. When the calibration starts the differential offset component is first calibrated for some time. Then the common offset component is calibrated for a few cycles. This alternating calibration of common and differential offset continues until the offsets of both residue amplifiers are below $\frac{LSB}{10}$. The calibration shown in figure 6.19 does not require any special calibration input signal, but operates with any regular or irregular signal applied to the A/D converter, as has been proven in section 4.5. The offset is not necessarily constant over time but can change due to temperature or supply changes. As a result, once the offset is reduced below the desired threshold, the calibration is not stopped but continuously monitors whether an offset appears again on the amplifiers, and if so it calibrates them.

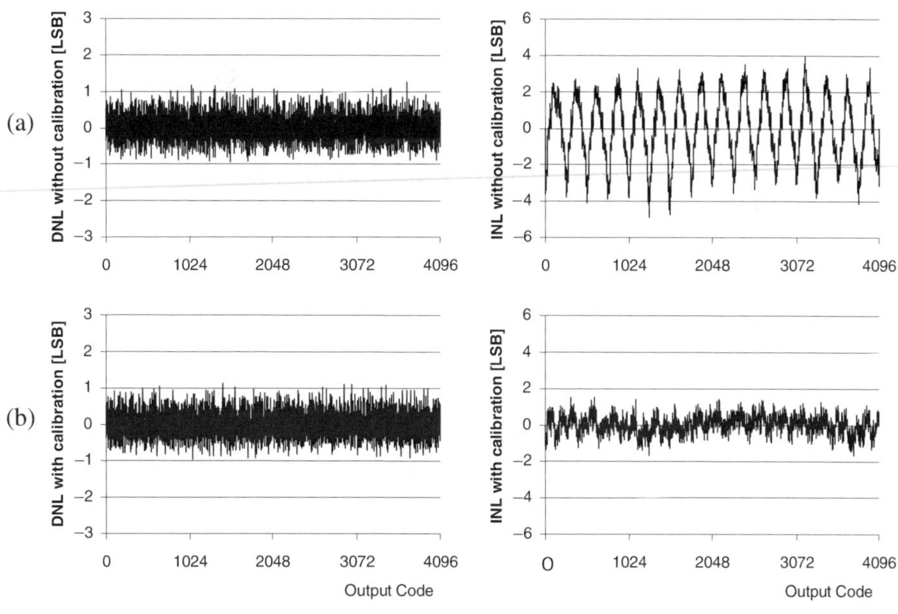

Figure 6.20: *Measured DNL and INL with $f_s = 54$ MSample/s and $f_{in} = 1.0$ MHz without calibration (a) and with calibration (b)*

Figure 6.20 shows the DNL and INL with (a) and without (b) calibration. The DNL is not improved with this algorithm, as it is already small due to the dual-residue signal switching (section 4.3.2). The upper peaks in INL shown in figure 6.20a are a measure of the offset on amplifier B (figure 6.9); the lower peaks are a measure of the offset on amplifier A. The DNL and INL after calibration of figure 6.20 could have been further improved by improving the pre-amp and comparator offset in the fine A/D converter.

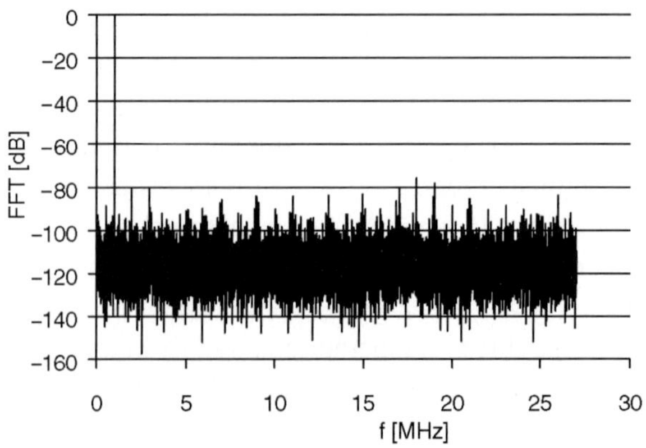

Figure 6.21: *Measured spectrum (with calibration) with $f_s = 54$ MSample/s and $f_{in} = 1.0$ MHz*

A measured spectrum is shown in figure 6.21. The measurement has been performed at 54 MSample/s and a signal frequency of 1.0 MHz. It can be seen that the harmonic performance is good due to the offset calibration of the residue amplifiers. All spurious components are below 75 dB.

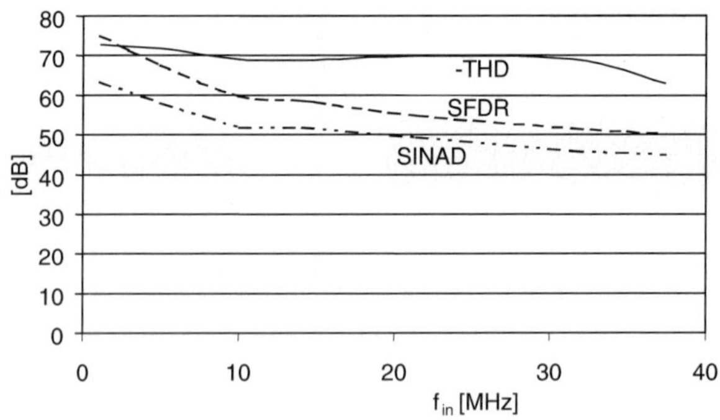

Figure 6.22: *Dynamic performance (with calibration) measured with $f_s = 54$ MSample/s*

The dynamic performance measured at 54 MSample/s in figure 6.22 shows good THD performance up to the Nyquist frequency of –70 dB. However, the overall performance decreases due to spurious components at $\frac{f_s}{3} \pm f_{in}$, which are non-harmonic spurs. The SFDR does not therefore reduce the THD performance, but reduces the signal-to-noise and distortion (SINAD). The components come from a frequency-dependent mismatch in the three interleaved sample-and-holds, which is most probably caused by a wiring capacitance mismatch in the analog input signal lines and clock signal lines [123]. Because the resolution bandwidth is limited by the interleaved sample-and-holds, the Figure-of-Merit calculated with equation 3.35 has a large value of 29,25 pJ/conv. When normalized to 1 V input voltage amplitude (equation 3.37), it equals: 94,77 pJV^2/conv. Table 6.2 summarizes the performance.

Table 6.2: *Summarized performance, f_s = 54 MSample/s and f_{in} = 1.0 MHz unless stated otherwise*

Technology	1P5M 0.25 μm CMOS
Resolution	12 bits
Supply voltage	2.5 Volt
Sample rate	54 MSample/s
Input range	1.8 V_{pp}, differential
SNR	64 dB
THD	−72 dB
SFDR	75 dB
SINAD	63 dB
INL	1.7 LSB
DNL	1.1 LSB
ERWB	4 MHz
Power Dissipation	295 mW
Area	1.0 mm^2

6.6 Discussion

In this design no special care has been taken to create an exact gain factor with the residue amplifiers. The dual-residue signal processing circumvents this requirement. The offsets of these residue amplifiers are reduced by the calibration algorithm. This reduces the INL from ± 5 LSB to ± 1.7 LSB. For the application in the GSM base-station signal processing chain this INL is sufficient, as the derived specification points (THD and SFDR for low signal frequencies) meet the requirements. After calibration the INL are most probably limited by the 8-bit folding A/D converter which generates the fine bits. This is illustrated in figure 6.20 by the repeating pattern in the INL. To improve this larger input devices of the fine A/D converter pre-amplifiers are required. This would increase the power-consumption. A better solution is to replace the folding A/D converter by a more power-efficient pipe-line A/D converter.

The SNR is 2 dB below the application specification from table 6.1. The thermal noise is dominantly generated in the sample-and-hold and in the residue amplifier stage. Further reduction of the thermal noise increases the power consumption. If it is assumed that the thermal noise of the S/H is dominant in the overall noise, then increasing the SNR by 2 dB means a factor of 1.6 power consumption (equation 2.12 and equation 2.15). This increases the overall power consumption to 365 mW. However, the SNR is also limited by the spurious components originating from the three-times interleaved sample-and-hold. At higher signal frequencies in particular this limits the SNR and the SFDR as well. The SFDR at low input signal frequency is sufficient but degrades at higher input signal frequencies. To increase the SNR and SFDR performance at higher input signal frequencies, further investigation into interleaving needs to be carried out [123]. The THD shows good performance over the whole Nyquist band compared to the specified THD in table 6.1.

The major limitation in the A/D converter realized is the limited matching between the three sample-and-hold circuits. This causes large spurious tones at $\frac{f_s}{3}$ and $\frac{f_s}{3} \pm f_{in}$. These spurious tones limit the resolution bandwidth to only 4 MHz. The Figure-of-Merit calculated with equation 3.35 therefore has a large value of 29,25 pJ/conv. If this is normalized to 1 V input voltage amplitude (equation 3.37), it equals 94,77 pJV^2/conv. When the resolution bandwidth is improved to the Nyquist frequency by improving, for example, the layout of the interleaved sample-and-holds, the $FoM_{V_{pp}}$ would be greatly improved to less than 16 pJV^2/conv. The power can be reduced by using a more power-efficient 8-bit pipe-line second stage, instead of a folding A/D converter, as is shown in figure 3.32. This does, however, increase the latency of the A/D converter.

6.7 Conclusions

The experiment described in this chapter has shown that the accuracy problems in CMOS A/D converters can be improved without impairing the bandwidth or requiring excessive power. The benefits of dual-residue signal processing in combination with MSCC have been verified experimentally. It has been shown that eventhough simple open-loop residue amplifiers with relatively small input devices have been used without special care being taken to ensure an exact gain factor, the MSCC reduces the offsets of the residue amplifiers to a sufficiently low level, as is demonstrated by the INL improvement in figure 6.20. The method presented does not require part of the sample period, thereby optimizing the sample rate. The limits of the method in terms of achievable accuracy form the subject of the next chapter.

Chapter 7

A low-power 16-bit three-step ADC for imaging applications

7.1 Introduction

The realizations in the previous chapters demonstrated the possibility of dual-residue signal processing in combination with offset compensation but were not optimized for power. The scope of this chapter is to design an A/D converter with a low power consumption and a high dynamic range (DR_{pp}) suitable for an imaging application. CMOS sensors (low-end) or CCD sensors (high-end) are used for digitizing images. The DR_{pp} defined as peak-to-peak signal to rms noise ratio of these sensors is around 60 dB and 75 dB respectively. Since the sensitivity of the human eye is only around 60 dB, it is not necessary to digitize the complete dynamic range of the image sensor. This means that an A/D converter with a accuracy of only 10 bits is required. Additional required processing is performed in the analog domain. This processing includes gain or compression. The gain is reduced by the digital processor, which detects when the signal from the sensor becomes too large. Compression can be used to increase the resolution in the dark part of the image, since the eye is more sensitive to the darker parts of the image. The block diagram of such a sensor signal conversion system is shown in figure 7.1a.

The digitization in figure 7.1a is performed with a 10-bit A/D converter. The DR_{pp} of this system is sufficient for low-end sensor systems. However, when

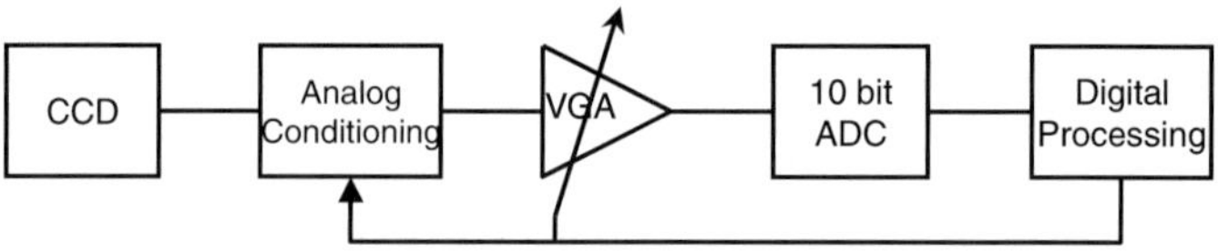

(a)

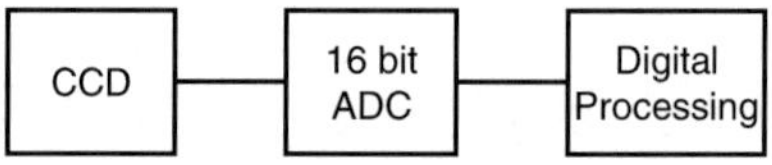

(b)

Figure 7.1: *Traditional moving image conversion system (a) and image conversion system with high dynamic range A/D converter (b)*

CCD sensors which have larger DR_{pp} are used, the analog processing shown in figure 7.1a can be shifted to the digital domain, making the required image processing much more flexible. The resulting chain from sensor to A/D converter is shown in figure 7.1b. In this case a 16-bit A/D converter is used to convert the full DR_{pp} of the CCD. All the image processing such as gain and compression is carried out in the digital domain. The main specifications for the 16-bit A/D converter are shown in table 7.1.

Table 7.1: *Requirements for high-end image conversion*

Technology	Digital CMOS without options
Sample Rate	30 MSample/s for Mega-Pixel CCD
DR_{pp} CCD	75 dB (in 5 MHz)
DR_{pp} ADC	84 dB (small input level in 5 MHz)
Resolution	16 bit
DNL	Strict monotonical (<1 LSB)
Full scale settling	<1%
THD	60 dB
INL	Up to 15 LSB tolerable
Power dissipation	<150 mW
Area	small area for embedding

Table 7.1 shows the requirements for a high-end image conversion system. Cost requirements demand that the A/D conversion function is embedded in a digital CMOS chip. This enables integration of the large digital signal processing and the analog functionality on one chip. The sample rate has to be 30 MSample/s

to accommodate the digitization of mega-pixel charge-coupled devices (CCDs) with 50 Hz frame rates. The state-of-the-art CCDs achieve 75 dB DR_{pp}. In order to limit the overall signal degradation in the system to 0.5 dB, the DR_{pp} of the A/D converter should be better than 84 dB in the 5 MHz video bandwidth. This high DR_{pp} has to be achieved for small input signal amplitudes. This corresponds to dark parts of an image for which the eye is most sensitive. The performance is allowed to degrade for large input signal amplitudes because the eye is much less sensitive to such signals. This DR_{pp} could be realized with a 13-bit A/D converter. However, a technique frequently used in imaging applications involves averaging a number of subsequent images to reduce the noise (at the cost of lower frame rates). With only 13 bits of resolution the quantization noise dominates the overall noise after averaging. This leads to unwanted artifacts in the picture. The specified resolution is therefore 16 bits.

The DNL should be within ± 1 LSB to guarantee monotonicity and non-missing codes. The performance for input signals with a large amplitude is less important since the perception of information in highly illuminated parts of a picture is less critical. This applies to the requirements in full-scale signal settling. The linearity parameters INL and THD are not critical in this imaging application. The color performance determines the minimum THD (and maximum INL).

The power consumption is very important for imaging applications, as the IC containing the A/D converter is close to the CCD and heating of the CCD increases the thermal noise.

The previous chapter has shown a realization of a two-step A/D converter with digital extraction and analog compensation of the residue amplifier offsets. This chapter extends this technique in a 16-bit three-step A/D converter which uses 2 separate and independent mixed-signal chopping and calibration (MSCC) blocks to calibrate the offset of 2 (the mid and fine) dual-residue amplifier stages [3]. At the same time, the overall power consumption must be reduced to the 150 mW allowed in this application. This experiment will show that this extreme accuracy can be achieved with a modest power consumption although some bandwidth reduction is inevitable.

7.2 Three-step architecture

The 16-bit three-step A/D converter architecture is shown in figure 7.2.

The three-step A/D converter is similar to the two-step A/D converter described in the previous chapter. However, using the two-step architecture for generating

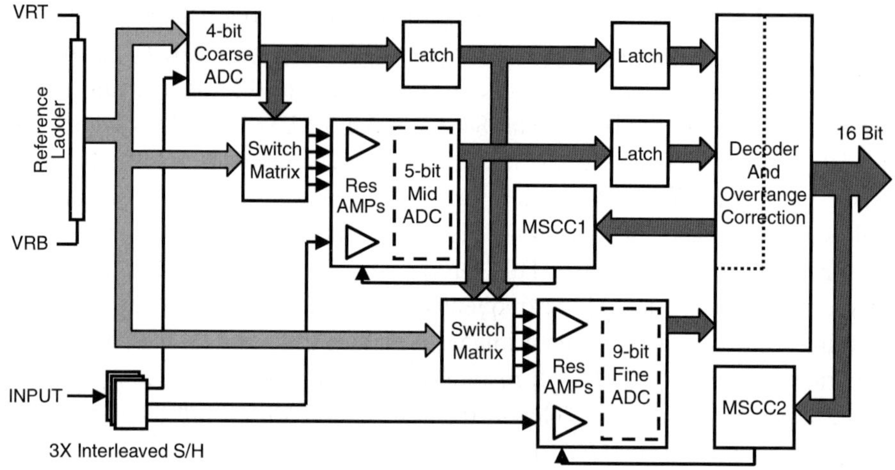

Figure 7.2: *Block diagram of the 16-bit three-step A/D converter*

16-bit output data would dramatically increase the accuracy requirements of the 2 quantization stages. An additional quantization step is therefore added to reduce the accuracy requirements of the quantization stages. The output signal from the CCD is sampled by a three-times interleaved [122] sample-and-hold (S/H), eliminating the need for re-sampling of the signal after each A/D converter stage. The required accuracy of the coarse A/D converter is limited to only 5 bits due to over-range in the next stage (section 3.4) [29]. This quantization result is used to select the references for the mid quantization in the next clock phase (similar to section 6.2.3). The selected references are combined with the held input signal in 2 dual-residue amplifiers, as explained in section 6.2.3, that are offset calibrated (section 4.5). The 5-bit mid A/D converter quantizes the output signals of these mid-residue amplifiers. The required accuracy of the mid A/D converter is limited to 6 bits, since the fine A/D converter also incorporates over-range (section 3.4) [29]. The outputs from both coarse and mid A/D converters are combined in order to select 4 references out of 256 levels for the fine quantization. These 4 references are combined with the sampled input signal in 2 dual-residue amplifiers (as explained in section 6.2.3), which are offset calibrated. These amplifiers realize a gain factor of 64 and consist of 2 cascaded amplifiers to improve speed [9]. Due to the dual-residue signal processing the absolute value of this gain is not critical (section 4.3.2) and therefore open-loop amplifiers can be used. The amplified residue signals are applied to a 9-bit fine A/D converter, which is realized as an 8 times folding A/D converter (similar to chapter 6 but with 8 times folding instead of 4 times) [21]. The overall A/D converter consists primarily of non-critical components, such as low-resolution quantizers, switches and open-loop amplifiers. The noise is mainly determined by the S/H and the fine residue amplifiers, while it is primarily the S/H and the static reference ladder [121] that determine the overall linearity.

7.2.1 Sample-and-hold

The input signal is sampled by a S/H. Since the signal from the CCD is a sampled signal, the coarse A/D converter can perform the coarse quantization at the same time. After the coarse quantization the input signal together with the selected mid reference signals have to be applied to the mid A/D converter after a sample period to perform the mid quantization. In combination with the previous coarse quantization the mid A/D converter selects the fine reference signals. The selected fine reference signals are combined with the input signal that is again held for one sample period and quantized in the fine A/D converter. To make the proper delayed input signals, a three-times interleaved S/H circuit is used, consisting of 3 identical sample-and-hold units, each sampling the input signal at one-out-of-three sample periods. The timing of the S/H is shown in figure 7.3.

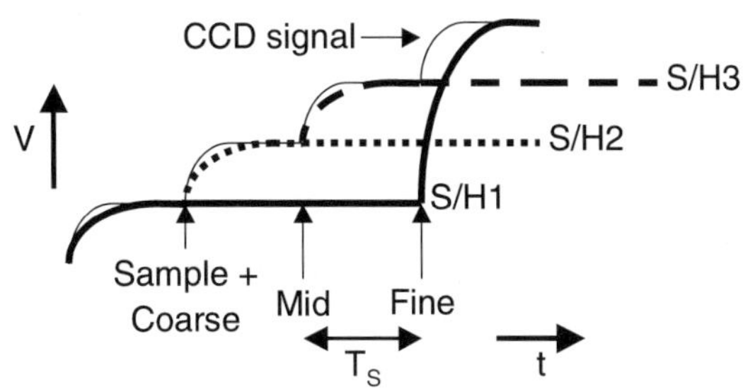

Figure 7.3: *Timing of the sample-and-hold circuit*

To reduce the power consumption, the analog bandwidth is limited since the CCD provides a signal that has already been sampled and the accuracy of large signal transitions is less critical. The matching of the three interleaved S/H circuits is critical. Static mismatch causes a component in the spectrum at $\frac{f_s}{3}$, while a dynamic mismatch causes components to appear at $3f_s/3 \pm f_{in}$ [123]. However, since the signal from the CCD is already sampled the dynamic mismatch is not critical. The static mismatch can be reduced by optimal switch sizes and careful layout.

7.2.2 Resolution distribution

The resolution can be distributed over the three steps in several combinations. The optimum distribution for minimum total intrinsic capacitance is when the resolutions of the three steps are equal (similar to section 3.18). This results in 6 coarse bits, 6 mid bits and 6 fine bits. The total number of bits of the coarse, mid and fine A/D converter amounts to more than 16 bits since over-range is employed in the mid and fine stage (section 3.4) [29]. This choice of resolutions requires 4 times 2048 switches and taps on the reference ladder for the fine reference selection. This results in a large parasitic capacitance connected to the reference ladder. To reduce the number of required switches the sum of the coarse and mid bits is reduced, which results in an increased number of fine bits. This design uses 4 coarse bits, 5 mid bits and 9 fine bits. This requires 4 times 16 switches to select the mid references and 4 times 256 switches for the selection of the fine references which results in a factor of 4 reduction in parasitic switch capacitance. The increased accuracy places demands on the fine A/D converter, which increases from 7 to 9 bits, and is resolved by using a folding A/D converter. The INL of a folding A/D converter depends mainly on the size of the input transistors of the folding amplifiers (section 3.3). However, because of the folding amplifier gain, the DNL is small. In the imaging application DNL is more important than INL. This allows for smaller folding amplifier input devices that reduce the intrinsic capacitance and therefore power.

7.2.3 Switching

The selection of the references for the mid conversion is made by translating the thermometer code of the coarse conversion to a 'one-high' code. This 'one-high' signal selects the 4 proper reference signals from the static reference ladder (see chapter 6).

Since the selection of the references for the fine conversion requires 1 out of 256 selection, special care has been taken. To avoid the need for an 8-bit binary to 'one-high' encoder, the coarse quantization selects 1 out of 16 switch blocks, in which the mid quantization selects the 4 proper references. This is similar to the matrix selection used in DACs [125]. Before the proper coarse selection can be made, the errors of the coarse quantization have to be corrected first. This is done by using the under and over-range indicator of the mid A/D conversion and is shown in figure 7.4.

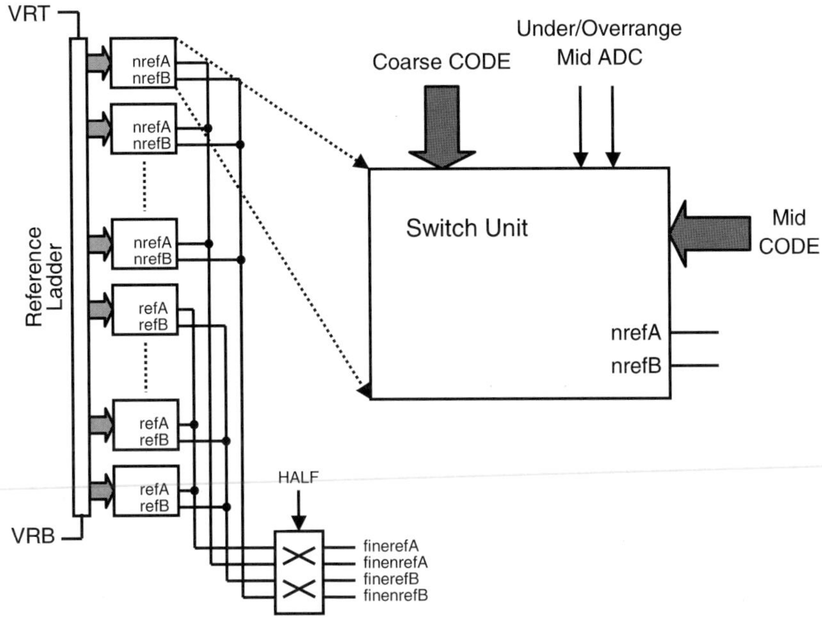

Figure 7.4: *Fine switch matrix with selection of the fine reference signals using matrix encoding and reference symmetry*

To reduce the number of switches required, the property of symmetry is used. When a certain fine subrange is selected, for example at the bottom of the range, references from the bottom and references from the top of the resistor ladder are selected since the input signal is differential. When the input signal is at the top of the range, the same references are selected but are applied inversely to the fine residue amplifiers. In this case the same reference selection is made, but the references are exchanged by a set of additional switches. This is also shown in figure 7.4. The inversion of the selection is indicated by the signal *HALF*. *HALF* is zero when the input signal is in the lower half of the input signal range and *HALF* is one when the input signal is in the upper half of the input signal range. The corresponding selected reference signals (before processing with the *HALF* signal) are shown in figure 7.5. For the sake of simplicity the under-range and over-range have been omitted from this figure.

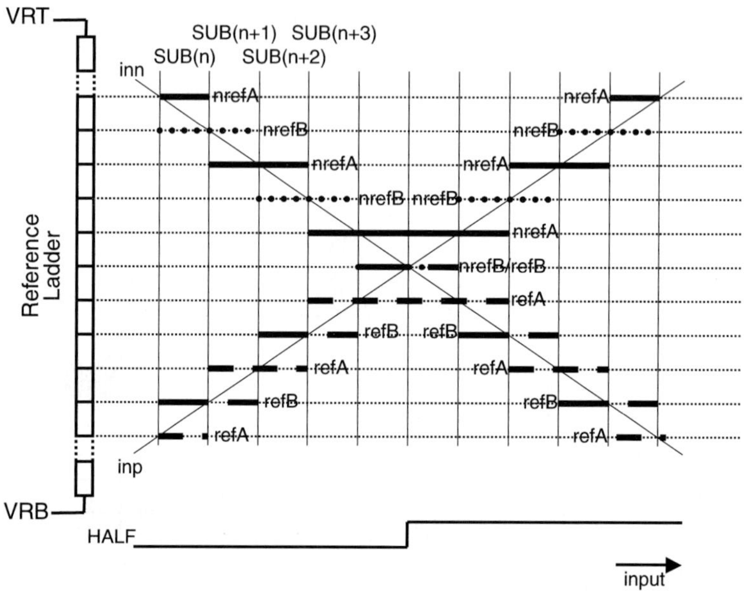

Figure 7.5: *Selected fine reference signals in the switch matrix before the inversion with the HALF signal (without over-range)*

As shown in figure 7.5, the references are selected in a similar way to that described in chapter 6. However, the selected signals are the same for the upper half and the lower half of the input range, which results in only half the number of selection switches being used.

7.3 Noise considerations

The S/H consists of three-times interleaved S/H units, as explained in section 7.2. Figure 7.6a and figure 7.6b show the sample and the hold phase, respectively, of such a flip-around [56, 33] S/H unit.

During the sample phase (figure 7.6a), the OPAMP acts as a virtual ground. The sampled noise in this case is given by equation 2.8, where C_{hold} is the hold capacitance and *NEF* is the noise excess factor [7] of the OPAMP. For more than 85 dB DR_{pp} and an input signal swing of 2 V peak-to-peak differential, a capacitance of 7.5 pF for C_{hold} in figure 7.6 is sufficient.

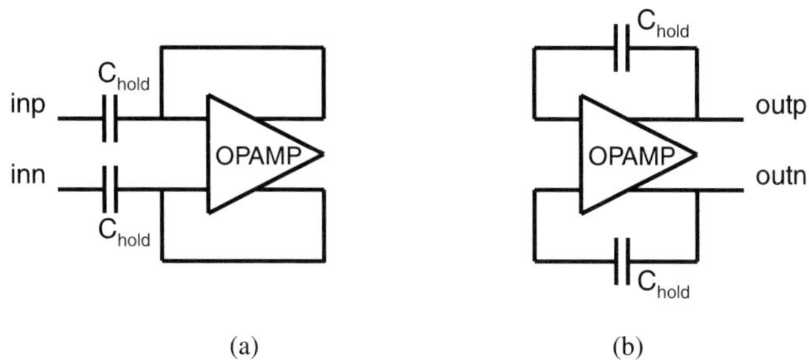

Figure 7.6: *The sample phase (a) and the hold phase (b) of a S/H unit*

The coarse and mid quantization do not determine the final quantization and are only instrumental for the fine reference selection. As a result, only the noise generated in the signal path to the fine comparators during the hold phase of the S/H is considered. During the hold phase (figure 7.6b) the hold capacitance is flipped around the OPAMP. During this phase the hold capacitance is not seen by the OPAMP as a load capacitance and therefore does not filter the noise generated by the OPAMP. The total noise power generated in the signal path during the hold phase is the sum of the noise generated by the OPAMP and the switches of the S/H, the reference ladder including switches, the residue amplifiers and the fine A/D converter. The effect of the latter is, however, reduced significantly by the gain in the residue amplifier stage. The noise generated during the hold phase of the S/H is sampled by the fine comparators. The gain in the signal path from the S/H up to the fine comparators reduces the accuracy requirements at the input of the fine comparators to less than 7 bit. The signal bandwidth at the comparator input is therefore limited with respect to the bandwidth of the S/H, resulting in a filtering of part of the noise coming from the S/H which is sampled by the fine comparators.

The accuracy requirements of the input stage of the S/H OPAMP and the residue amplifiers with respect to noise are similar. The noise contributions of these components are therefore similar for optimum power consumption. The noise contribution of the reference ladder is determined by the total resistance of the ladder. Reducing the ladder resistance increases the current through the ladder. Increasing the width of the selection switches reduces their noise contribution but increases the capacitive load determined by the switches, reducing the settling speed of the references. The noise contribution of each of the components described has been designed for minimum power consumption. The contribution to the DR_{pp} of each component is shown in table 7.2.

Table 7.2: *Contribution of the thermal noise to the DR_{pp} of the different A/D converter blocks during the sample and hold phase*

Phase	Sample	Hold				
Block	SHA	SHA	Ref ladder	Res Amp A	Res Amp B	Fine ADC
DR_{pp}	93.9 dB	98.31 dB	100.2 dB	107.22 dB	99.48 dB	121.54 dB
Sub total	93.9 dB	93.0 dB (total from noise simulation)				
Total	90.4 dB					

As can be seen, the contribution of both residue amplifiers is not equal. This is dependent on the position of the zero-crossing in the fine A/D converter. Close to the beginning or the end of the fine range the noise is dominated by one amplifier, while in the middle of the fine range both residue amplifiers contribute equally to the noise. The total DR_{pp} of 93.0 dB during the hold phase in table 7.2 is not calculated from the sum of the contributions of each block, but is the simulated total value. The sum of the noise during the hold phase in table 7.2 results in a DR_{pp} of 94.3 dB. This means that the remaining blocks of the A/D converter generate noise resulting in a DR_{pp} of 98.8 dB. These remaining blocks include the chopping circuits and biasing. The total DR_{pp} determined by thermal noise during the sample and the hold phase equals 90.4 dB. This gives some margin in the design with respect to other sources of noise, such as reduction due to comparator offset errors or substrate noise injection.

7.4 Mixed-signal chopping and calibration

7.4.1 Mid and fine residue amplifier stage calibration

The offsets of the mid residue amplifiers need to be below 128 LSB at 16-bit accuracy to guarantee proper fine sub-range selection, while the offsets of the fine residue amplifiers need to be below 1 LSB as this determines the accuracy of the ADC. These low offset values are obtained by using MSCC, with digital offset extraction and analog compensation [58]. The offset extraction of both mid and fine residue amplifier stages is performed simultaneously.

The coarse A/D converter quantizes the input signal and determines in which mid subrange the input signal is located. The corresponding references are selected by the mid switch matrix and applied together with the input signal to the mid residue amplifiers. By chopping the input signals of these residue amplifiers, their offset is detected using the digital output of the mid A/D converter. Chopping of the fine residue amplifiers is applied after the mid quantization is performed, as can be seen in figure 7.2. The fine residue amplifier chopping does not therefore have any effect on the mid quantization and the extraction of the mid residue amplifier offset is performed independently of the chopping of the fine residue amplifiers.

Since the mid residue amplifier chopping is performed before the fine quantization, this can potentially influence the fine residue amplifier offset extraction. Below it is explained how the mid residue amplifier offset extraction and the fine residue amplifier offset extraction do not influence each other.

The mid quantization determines the selected fine reference signals and chopping of the mid residue amplifier with offset may therefore lead to a different fine subrange (and fine reference signals) selection. The result is that for a certain input signal two different fine-residue signals can be applied to the fine A/D converter. This is illustrated in figure 7.7a.

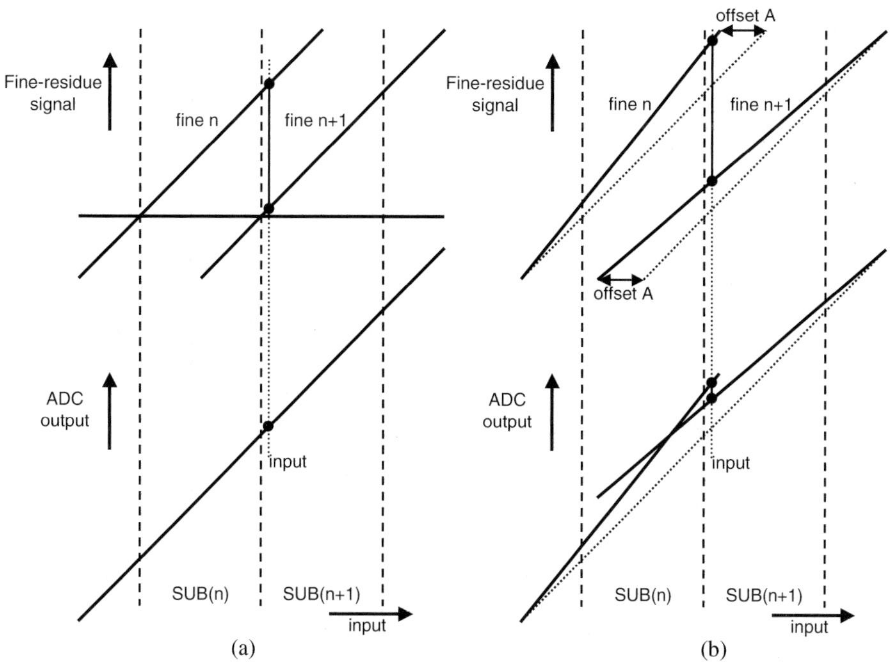

Figure 7.7: *Fine residue signal selection and A/D converter output without (a) and with (b) fine residue amplifier offset*

The residue signal $fine(n)$ in figure 7.7a is the result of the selection of $SUB(n)$ by the mid A/D converter, while residue signal $fine(n+1)$ is the result of the selection of $SUB(n+1)$. Because there is overlap (redundancy) between the subranges $SUB(n)$ and $SUB(n+1)$, the same A/D converter output code is generated, independently of the result of the mid A/D converter.

When there is also an offset present at the input of the fine residue amplifiers, the corresponding residue signals of two subsequent subranges are shown in figure 7.7b. In this case only amplifier A has an offset of $offsetA$. When, due to mid residue amplifier offset and chopping, two different subranges $SUB(n)$ and $SUB(n+1)$ are selected for a certain input signal, this results in two different fine residue signals $fine(n)$ and $fine(n+1)$. However, due to fine residue amplifier offset the overlap is not perfect, as is shown at the bottom of figure 7.7b. The total A/D converter therefore has different output codes for constant input signal, illustrated with the black dots in figure 7.7b. As can be seen, the chopping of the mid residue amplifiers changes the output of the A/D converter. However, the contribution of the offset of the fine residue amplifiers to the A/D converter output, with respect to the ideal output, still results in a positive error. The fine residue amplifier offset extraction is therefore not disturbed, since the offset extraction only detects the sign of the offset.

7.4.2 Quick calibration

The MSCC only updates the DAC values after an integration period, which is sufficiently long for reliable and accurate offset extraction. The MSCC therefore takes over 10^6 samples to settle. For test purposes a quick foreground calibration option is integrated during fabrication for each MSCC; this can also be used at power-up for a quick initial calibration. During quick calibration mode the inputs of the residue amplifiers are short-circuited and the digital extraction block applies a binary search [87] for the proper compensation DAC codes. For the comparison during the binary search, the ladder buffers, pre-amplifiers and comparators of the normal operation mode are used. The quick calibration system of the residue amplifiers and the corresponding D/A converter signals of one calibration cycle are shown in figure 7.8.

Since the D/A converters for the mid and fine compensation are 9 and 11 bits, respectively, the quick calibration cycles take 9 and 11 clock cycles.

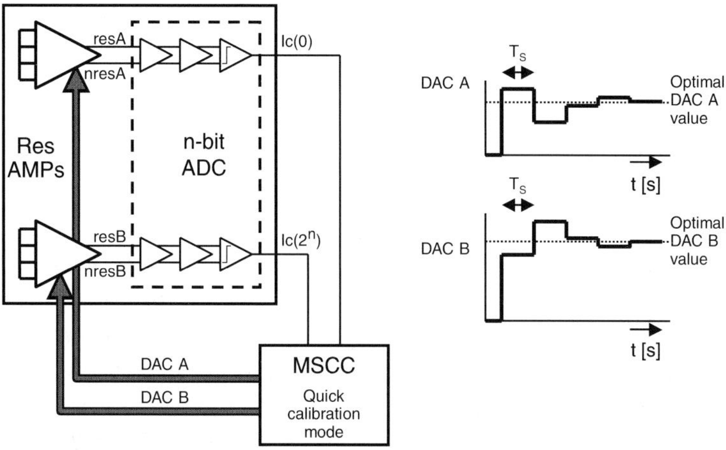

Figure 7.8: *Quick calibration system with calibration cycle example*

7.5 Supply voltages

Most CMOS technologies offer dual gate oxide. For this design 0.18 μm CMOS technology with 2 values for the gate oxide thickness was used. The (effective) oxide thickness determines the permitted supply voltage (thin oxide for 1.8 V devices and thick oxide for 3.3 V devices) and therefore the achievable signal swing. Figure 2.7 shows the relative dependence of the power as a function of the (effective) oxide thickness. For noise- and matching-limited designs the capacitance increases by $\frac{1}{t_{ox}^2}$ and $\frac{1}{t_{ox}}$ respectively. To compare the power, it is assumed that the bandwidth remains constant and V_{gt} scales linearly with V_{dd}. As is shown in figure 2.7, the power for a noise limited design is initially constant as a function of t_{ox}. However, since the relation between I_{ds} and V_{gt} is not linear for small V_{gt}, for small t_{ox} the power becomes *inversely* proportional to $\sqrt{t_{ox}}$ for a noise-limited design (which is also shown in figure 2.7). For a matching-limited design, the power is initially proportional to t_{ox}, however for small V_{gt}, the power becomes proportional to $\sqrt{t_{ox}}$. This is illustrated in figure 2.7. As this ADC has both noise-limited and matching-limited parts, the optimum power dissipation is exploited by using two different gate oxides. Figure 7.9 shows the block diagram of the A/D converter, with the supplies per block.

The capacitance of the S/H is determined by the noise requirements and is designed with the 3.3 V thick oxide transistors. The quantizing parts, which are matching limited, and the digital circuitry are designed with 1.8 V thin oxide transistors. The power consumption of the 3.3 V and the 1.8 V part are now similar.

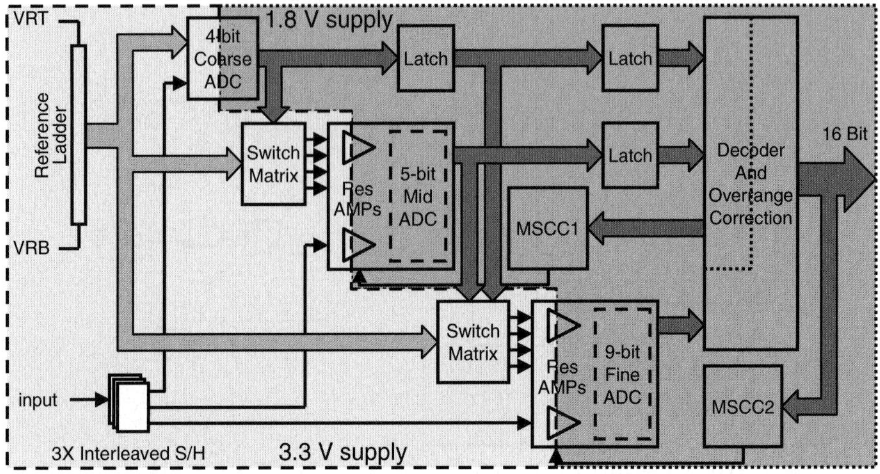

Figure 7.9: *Block diagram with two separate supplies*

Changing the 3.3 V supply into a 1.8 V supply would decrease the allowable signal swing by a factor of 1.8. This would increase the required power consumption by a factor of 1.3, since the capacitance is increased by this factor and in the OPAMP of the S/H input devices with small V_{gt} are used.

7.6 Experimental results

A die photograph of the three-step A/D converter is shown in figure 7.10.

The A/D converter is fabricated in standard single poly, five metal 0.18 μm CMOS and the core area is 1.4 mm^2. The ADC operates at 1.8 V and 3.3 V supply voltages and dissipates 141 mW (without output buffers). The relative power consumption of each block is shown in figure 7.11.

The power consumption of the fine residue amplifiers is relatively large. This is caused by the fact that these amplifiers have a large gain ($\approx$64) in combination with a large signal bandwidth. The fine A/D converter power consumption is relatively high since 9 bits are resolved in one step by a folding A/D converter.

For imaging applications, and in particular for a small input signal, the DR_{pp} is very important. Figure 7.12 shows the DR_{pp} of the A/D converter, which is the

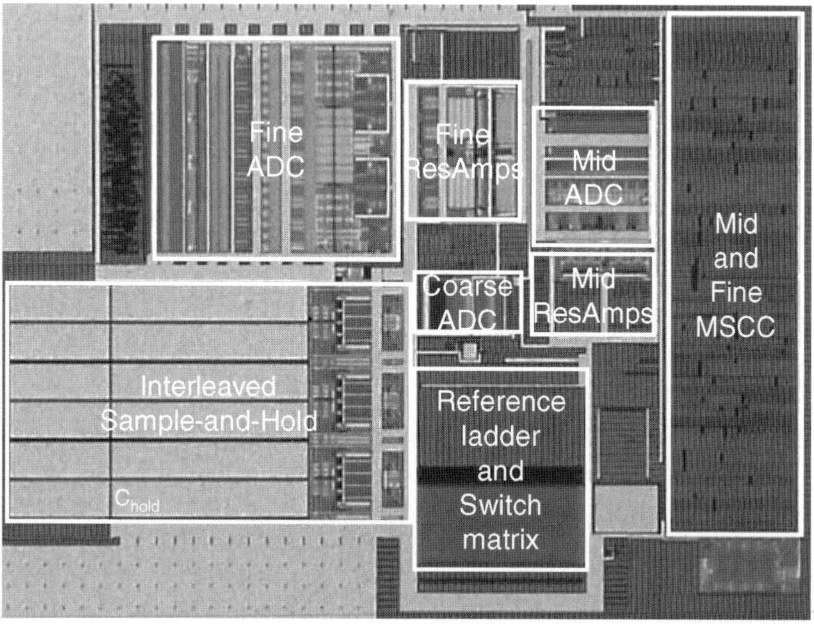

Figure 7.10: *Die photograph of the three-step 16-bit A/D converter*

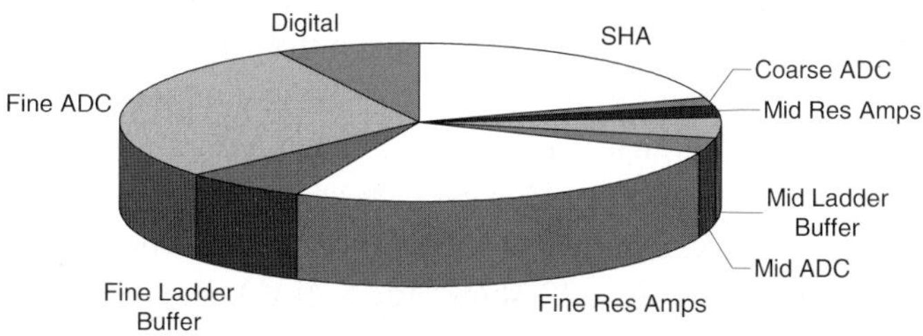

Figure 7.11: *The relative power consumption of each block of the A/D converter*

peak-to-peak signal to rms noise ratio for different input signal levels as a function of the sample rate. The measurements have been performed with an input signal frequency of 1 MHz.

When no input signal is applied, the DR_{pp} of the A/D converter is calculated by dividing the peak-to-peak output code (which is 2^{16}) by the rms of the noise measured at the output of the A/D converter. This DR_{pp} is above 88 dB up to 30MSample/s, which is close to the calculated value from table 7.2. For low input signal levels, the DR_{pp} is above 85 dB for sample rates up to 30 MSample/s. For

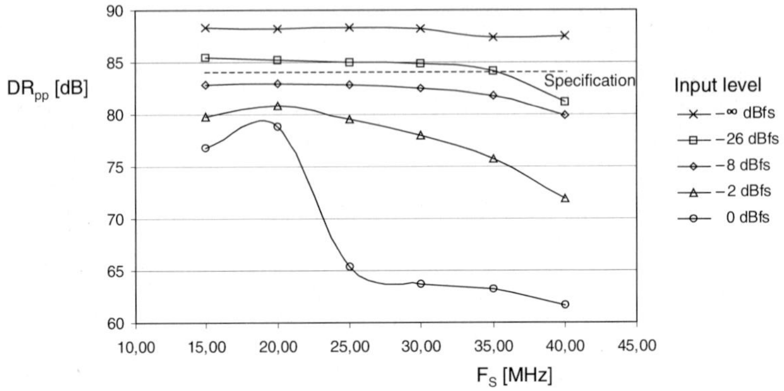

Figure 7.12: *DR_{pp} for different input signal levels as a function of the sample rate (f_{in} = 1 MHz)*

larger input levels, the DR_{pp} degrades further. A major artifact which causes this has been found in the fine-residue stage. For large input signal amplitude (and relatively high signal frequency), the fine reference change is large from sample to sample. These large jumps in both references and sample-and-hold signal incidentally cause the output stage of the fine-residue amplifier stage to go out of saturation. Recovery from this requires too much time with respect to the highest sample rates, thus degrading the performance. This will be solved in a subsequent version. A second effect is that the switching of the reference signals causes the reference ladder to be disturbed. However, in an imaging application the large signal behavior is less important, since a performance reduction for high light intensities is less visible. Figure 7.13 shows the SINAD as a function of the input signal level for different sample rates (f_{in} = 1 MHz).

As can be seen in figure 7.13, the SINAD increases as a function of the input level. For high input levels, the SINAD reaches a maximum before the input level is at maximum. This is also limited by the fine-residue amplifier stage and the limited settling speed of the ladder. This is confirmed by the fact that the input level where the SINAD is at maximum decreases for increasing sample rates. The DNL and INL plots, obtained for a low-frequency input signal and sampled at 15 MSample/s, are shown in figure 7.14 and figure 7.15. The number of points that are used to calculate the DNL and INL is 32000000.

As is shown in figure 7.12, the large signal performance at a sample rate of 30 MSample/s is limited. Since this would also limit the DNL and INL measurement, a sample rate of 15 MSample/s is used to measure the DNL and INL. The

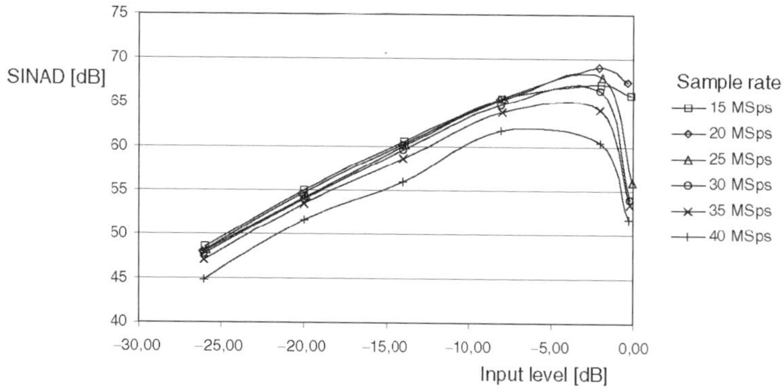

Figure 7.13: *SINAD as a function of the input signal level for different sample rates (f_{in} = 1 MHz)*

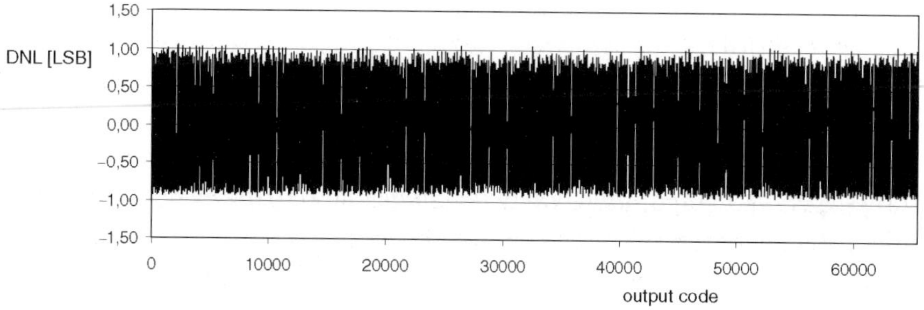

Figure 7.14: *DNL measured with a signal frequency of 230 Hz and 15 MSample/s*

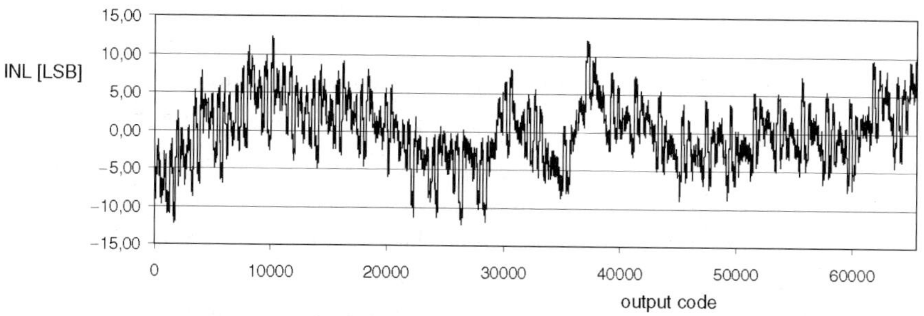

Figure 7.15: *INL measured with a signal frequency of 230 Hz and 15 MSample/s*

A/D converter has no missing codes since the DNL remains just above –1 LSB. The incidental peaks in DNL above +1 LSB are most probably caused by thermal noise. The INL is between –13 LSB and +13 LSB, which is sufficient for the application. In order to calculate the *FoM* of this converter, the resolution bandwidth has to be determined. The measurement of the SINAD as a function of the signal frequency for different signal levels is shown in figure 7.16.

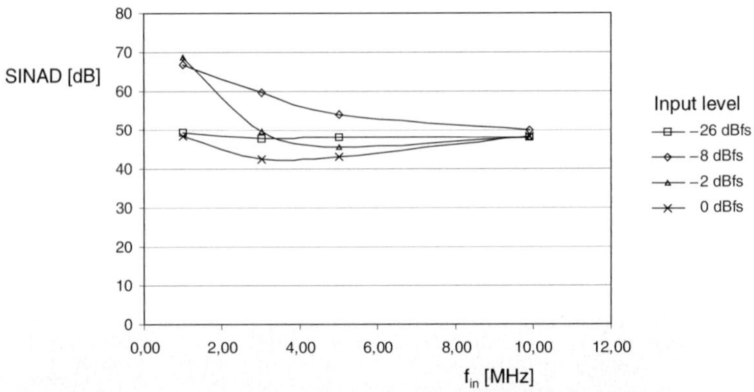

Figure 7.16: *SINAD as a function of the input signal frequency for different signal levels* (f_S = *20 MSample/s*)

Since the resolution bandwidth of this converter is limited to ≈2 MHz, the *FoM* is 19.7 pJ/conv. When the *FoM* is corrected for the input signal swing of 2 V peak-to-peak, the resulting $FoM_{V_{pp}}$ equals 79.1 pJV^2/conv. The performance is summarized in table 6.2.

Table 7.3: *Summarized performance*

Technology	5M1P 0.18 μm CMOS
Resolution	16 bit
Sample Rate	30 MSample/s
Input range	2 V_{pp}, differential
DR_{pp} ADC	88 dB (no input signal)
DR_{pp} ADC	77 dB (input signal–2 dBfs)
DNL	–1 LSB <DNL <1.1 LSB (no missing codes)
INL	–13 LSB <INL <13 LSB
SINAD (max)	68 dB
Power dissipation	141 mW
Area	1.4 mm^2

An A/D converter [126] for a similar application has shown a smaller sample frequency of 20 MSample/s and a larger power consumption of 300 mW (which includes the correlated double sampling circuit (CDS) and a programmable gain amplifier).

7.7 Discussion

This realization shows that the MSCC can be used in a 'more-step' configuration and is not limited to a two-step architecture. Together with dual-residue signal processing the use of MSCC relaxes the demands on the inter-stage residue amplifiers. The achieved DR_{pp} is 88 dB according to the specification. The DNL shows that the converter is monotonic, since the DNL is always larger than –1 LSB. Since the rms of the thermal noise voltage is about 2.5 LSB, a large number of points is used in the DNL calculation. However, this may still incidentally cause the DNL to be larger than 1 LSB.

The generation of 16 bits requires that 2^{16} levels are distinguished. To limit the number of comparators this is done in three steps. In this design, this does, however, still require 256 subranges to be selected from a reference ladder. This causes a large capacitive load of the reference selection switches to be connected to the reference ladders, which limits the achievable speed and input signal frequency. However, the sampling speed is sufficient for the imaging application and the performance for high input signal frequencies is not important. The speed could be improved by reducing the number of selected references by employing re-sampling of the analog signal after the gain stages. This would result in a more pipe-lined structure.

Because of the optimizations with respect to the input signal properties, such as high performance for small input signal levels and low performance for high input signal levels, the power is 141 mW. The large signal amplitude and high signal frequency performance is limited by slow recovery from over-drive at the output stage of the fine residue amplifier stage in this version of the 16-bit A/D converter. In the next version the output stage of the residue amplifier will be improved, increasing the resolution bandwidth. This will improve both the realized *FoM* and $FoM_{V_{pp}}$ since the improvement does not increase the power consumption.

7.8 Conclusions

The experiment in this chapter has shown that dual-residue signal processing with MSCC enables the design of a low-power A/D converter for imaging applications. The quantization is split up into three steps, reducing the accuracy requirements and power of the sub A/D converters. Due to the dual-residue signal processing and MSCC the A/D converter has no missing codes at 16-bit level even with the use of residue amplifiers with relatively small input devices and without an accurate gain factor. This is demonstrated in the DNL measurement. The DR_{pp} achieved is 88 dB in 15 MHz bandwidth for small input signal amplitudes at a power consumption of 141 mW. The performance degrades for larger input signal amplitudes, which is no limitation for the imaging application.

In addition to repairing the output stage of the fine residue amplifier stage in this version, an increase in the performance for larger input signal amplitudes and frequencies would require more power. The selection of the references in particular puts a limit on the high-frequency behavior. This A/D converter has been optimized for sampled input signals. When applying non-sampled signals the interleaved sample-and-hold probably has to be extended to a four-times interleaved sample-and-hold.

Chapter 8

Conclusions

Analog-to-digital conversion is the key technology for embedded signal processing. Bandwidth, accuracy and power are the main parameters that determine A/D converter performance. This book investigates calibration techniques to increase the accuracy of A/D converters while maintaining an acceptable power consumption for consumer applications.

Chapter 2 discusses the relation between accuracy and capacitance. Given the technology, the minimum required capacitances in an A/D converter are determined for both the noise and the matching requirements. The resulting capacitances derived from both the noise and the matching demands are quadratically dependent on the required accuracy. The maximum of these capacitances determines the required capacitance and is A/D-converter-architecure dependent. This required capacitance needs to be charged with the respective signal with sufficient accuracy within a certain time period, which is a measure for the power consumption. The power consumption of a circuit is proportional to the load capacitance and inversely proportional to the available settling time, if the parasitic capacitance is smaller than the aforementioned load capacitance. When the available settling time approaches its lower limit, which means that the parasitic capacitance reaches the load capacitance, the power consumption is no longer inversely proportional to the available settling time, but increases dramatically.

The errors caused by matching errors can be reduced by using calibration techniques. The capacitance resulting from matching requirements can therefore be decreased, increasing the power efficiency of the A/D converter. However, since noise is a random process, it cannot be calibrated and the capacitance determined by the noise requirements is the lower limit and cannot be reduced by calibration.

It is beneficial to use the most advanced technology for A/D converter designs, which are matching-limited. The effect of the decreasing power supply voltage reduces the allowed signal swing. However, the matching parameter A_{V_T} decreases as well. For noise-limited designs, the input signal amplitude should be as large as possible, which demands for older technologies with larger supply voltages. The maximum speed is achieved in the most advanced technologies, where the ratio between g_m and the parasitic capacitance generally increases.

The choice for an A/D-converter architecture allows a trade-off to be made between accuracy, speed and power. For low sample rates the architectures that reduce the number of necessary accurate devices are much more power-efficient than the parallel architectures. This is especially the case for high accuracy, where the required capacitance becomes impractical. However, high-sample-rate parallel architectures are more power-efficient for low accuracy, since the conversion is done in only one step. Practical realizations confirm this.

Higher-accuracy A/D conversion requires multi-stage architectures. When more bits in the first stage are applied, the accuracy in the least significant stages becomes less critical. Therefore, generating more bits in the first stage(s) reduces the accuracy requirements to only those of the first stage(s). This is in contrast to the parallel architectures where all the decision blocks need high accuracy.

Even in multi-stage architectures straightforwardly increasing the accuracy would lead to high power consumption. Therefore, designing A/D converters for high accuracy requires some form of calibration. Chapter 4 analyzes enhancement techniques for the two-step architecture to increase the accuracy without increasing the power consumption.

In multi-stage architectures gain errors are the dominant accuracy-limiting factors. Chapter 4 introduces the dual-residue signal-processing technique to circumvent these errors. When improved switching is used, continuity and monotonicity are guaranteed. The difference in offset of the two dual-residue amplifiers determines the INL of the overall A/D converter. For high accuracy this means that the input devices of the amplifier need to be large to ensure a sufficiently low offset voltage. To reduce the input capacitance of the residue amplifiers, calibration is applied.

The mixed-signal chopping and calibration technique (MSCC) enables offset extraction from the output data in the digital domain and compensation in the analog domain, resulting in an A/D converter with almost perfect analog circuitry. This technique uses only very few additional non-critical analog-components. It does not require additional calibration time and operates under normal conversion conditions. Therefore, the calibration can be performed continuously without interfering with the normal operation of the A/D converter. The signal processing capability of the digital CMOS technology is used to make large time-constants enabling highly accurate offset extraction. In contrast to other calibration techniques, the MSCC technique requires no additional post-processing of the digital output data and does not impair any timing parameter. The only overhead consists of some regular updating of the integrators.

The first experiment in chapter 5 shows the dual-residue signal processing in combination with analog offset calibration. Benchmarked with realizations in literature, the power consumption in this experiment is still too high, but the performance meets the requirements for CVBS video signal processing, which is used as a vehicle. The limitation of the calibration implementation in the first experiment is the result of the combination of A/D conversion and offset cancellation in one clock cycle.

The benefits of dual-residue signal processing in combination with MSCC is verified in the second experiment in chapter 6. It is shown that the accuracy problems in CMOS A/D converters are improved without impairing the sample rate, as was the case in the first experiment, or requiring excessive power. Even though simple open-loop residue amplifiers with relatively small input devices have been used, without special care for an exact gain factor, the MSCC reduces the offset voltages of the residue amplifiers to a sufficiently low level.

The experiment in chapter 7 shows that dual-residue signal processing with MSCC enables the design of a low-power A/D converter for imaging applications. The quantization is split-up into three-steps, reducing the accuracy requirements and power consumption of the sub A/D converters. Due to the dual-residue signal processing and MSCC the A/D converter has no missing codes at 16-bit level, even with the use of residue amplifiers with relatively small input devices and without an accurate gain factor. The performance degrades for larger input signal amplitudes, which is no limitation for the imaging application. Increasing the performance for larger input signal amplitudes and frequencies would require more power.

This book shows that accuracy in high-speed A/D converters can be optimized without sample rate or power penalties. Moreover, there is no need for extensive post-processing of the output data. The methods developed in this book seem applicable for other circuit applications as well.

Appendix A

Static and dynamic accuracy requirements

In this appendix the effect of static and dynamic errors on the performance degradation are calculated. With this result a maximum allowable error in the quantization is derived to limit the performance degradation to an acceptable level.

A.1 Static error requirments

To calculate the maximum allowable static errors, the effect of random errors on the transfer curve of the A/D converter (INL) is calculated. Since the effect of random errors on the transfer curve is similar to the effect of quantization noise on the transfer curve of the A/D converter, the noise power representing the random errors can be added to the error power caused by quantization (similar to [1]):

$$SNR_{Q+error} = 10\log\left(\frac{\frac{3}{2}2^{2N}}{1+12\sigma_{error}^2}\right), \qquad (A.1)$$

where σ is the standard deviation of the random errors and N is the number of bits. For an A/D converter this is equal to the standard deviation of the INL of each individual code. Subtracting the quantization noise from equation A.1 results in the reduction of SNR caused by INL errors:

$$SNR_{red} = -10\log\left(1 + 12\sigma^2_{error}\right) \tag{A.2}$$

When these INL errors are random errors with a gaussian distribution, the relation between σ of the errors and the maximum INL is given by:

$$\sigma^2_{error} = \frac{INL^2}{\alpha^2}, \tag{A.3}$$

where α is the threshold value for the stochastic variable of the probability that the maximum random error remains within the limits (–INL, +INL) is acceptable [1]. Depending on the number of critical components, α is in the range of three and five. In chapter 3 it is derived that for the scope of this book it is sufficient to set α to the constant value 4. Combining equation A.2 and equation A.3 results in the reduction of SNR caused by random gaussian distributed INL errors:

$$SNR_{red} = -10\log\left(1 + 12\frac{INL^2}{\alpha^2}\right) \tag{A.4}$$

To calculate the reduction in ENOB as function of the INL, equation A.4 is then rewritten to:

$$ENOB_{red} = \frac{-\log\left(1 + 12\frac{INL^2}{\alpha^2}\right)}{2\log 2} \tag{A.5}$$

The reduction in ENOB for a certain maximum INL is dependant on the distribution (shape) of the INL errors. The worst case situation occurs when the each individual INL error is equal to +INL or –INL. In that case, $\sigma^2_{error} = INL^2$ and equation A.5 changes into:

$$ENOB_{MAX\ red} = \frac{-\log\left(1 + 12\,INL^2\right)}{2\log 2} \tag{A.6}$$

Figure A.1 shows both $ENOB_{red}$ and $ENOB_{MAX\ red}$ as function of the INL. Also the result of a Monte Carlo analysis is shown. In this analysis, the INL is randomly determined. The x-axis shows the maximum INL, while the y-axis shows the reduction in ENOB. In the calculation of $ENOB_{red}$, the value for α is 3, which means that the INL used in figure A.1 is the 3σ value of the distribution of the INL.

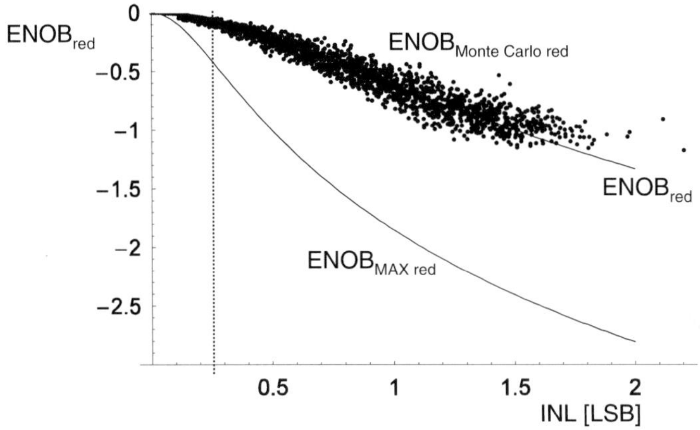

Figure A.1: *Reduction in ENOB for: a random INL (with $\alpha = 3$), only value of +INL or −INL, and a Monte Carlo simulation, as function of INL*

Figure A.1 shows that when the INL is less then ±0.25 LSB the overall performance decreases at maximum with 0.4 ENOB (or 2.4 dB). This is the case when the INL-pattern of the A/D converter consists of only the values +0.25 LSB and −0.25 LSB (on the curve $ENOB_{MAX\ red}$). In the remainder of this book the maximum loss of 0.4 ENOB is used in the calculations as allowed performance degradation. This means that all errors due to the signal processing such as settling or mismatch errors should remain within ±0.25 LSB. In practice the INL-pattern will also contain values ±0.25 LSB, which means the practical preformance degradation is in the order of 0.2 ENOB (on the curve $ENOB_{Monte\ Carlo\ red}$).

A.2 Dynamic error requirements

To keep the SNR reduction due to thermal noise (equation 2.12) sufficient low, in this book it is assumed that the signal-to-thermal-noise ratio is 6 dB better than the signal-to-quantization-noise ratio. This results in an overall SNR reduction of 1 dB, which is comparable to the reduction in SNR due to static INL errors, when the INL remains within ±0.25 LSB.

References

[1] M. Pelgrom, J. van Rens, M. Vertregt, and M. Dijkstra, "A 25-MS/s 8-bit CMOS A/D converter for embedded application," *IEEE J. Solid-State Circuits*, vol. 29, pp. 879–886, 1994.

[2] J. van Lammeren, *The design of low-cost one-chip TV systems*. PhD thesis, Universiteit Twente, 2000.

[3] C. Mangelsdorf, H. Malik, S. Lee, S. Hisano, and M. Martin, "A two-residue architecture for multistage ADC's," *ISSCC Digest of Technical Papers*, pp. 64–65, 1993.

[4] M. Pelgrom, A. Duinmaijer, and A. Welbers, "Matching properties of MOS transistors," *IEEE J. Solid-State Circuits*, vol. 24, pp. 1433–1440, 1989.

[5] M. Pelgrom, H. Tuinhout, and M. Vertregt, "Transistor matching in analog applications," *IEDM Technical Digest*, pp. 34.1.1–34.1.4, 1998.

[6] A. Scholten, H. Tromp, L. Tiemeijer, R. van Langevelde, R. Havens, P. de Vreede, R. Roes, P. Woerlee, A. Montree, and D. Klaassen, "Accurate thermal noise model for deep-submicron CMOS," *IEDM Technical Digest*, pp. 155–158, 1999.

[7] K. Laker and W. Sansen, *Design of analog integrated circuits and systems*. New York: McGraw-Hill, 1994.

[8] M. Vertregt and P. Scholtens, "Scalable high speed analog circuit design," in *Analog circuit design* (J. Huijsing, M. Steyaert, and A. van Roermund, eds.), pp. 3–21, Kluwer Academic Publishers, 2004.

[9] B. Razavi, *Design of analog CMOS integrated circuits*. New York: McGraw-Hill, 2001.

[10] D. R. Breuer, "High-speed A/D converter monolithic techniques," *ISSCC Digest of Technical Papers*, pp. 146–147, 1972.

[11] R. A. Nordstrom, "High-speed integrated A/D converter," *ISSCC Digest of Technical Papers*, pp. 150–151, 1976.

[12] J. G. Peterson, "A monolithic video A/D converter," *IEEE J. Solid-State Circuits*, vol. 6, pp. 932–937, 1979.

[13] G. Geelen, "A 6b 1.1GSample/s CMOS A/D converter," *ISSCC Digest of Technical Papers*, pp. 128–129, 2001.

[14] P. Scholtens and M. Vertregt, "A 6-bit 1.6-GS/s flash ADC in 0.18-μm CMOS using averaging termination," *IEEE J. Solid-State Circuits*, vol. 37, pp. 1599–1609, 2002.

[15] C. Lane, "A 10-bit 60 MSps flash ADC," *Proc. BTCM*, pp. 44–47, 1989.

[16] A. Venes and R. van de Plassche, "An 80-MHz, 80-mW, 8-b CMOS folding A/D converter with distributed track-and-hold preprocessing," *IEEE J. Solid-State Circuits*, vol. 31, pp. 1846–1853, 1996.

[17] K. Kattmann and J. Barrow, "A technique for reducing differential nonlinearity errors in flash A/D converters," *ISSCC Digest of Technical Papers*, pp. 170–171, 1991.

[18] K. Kusumoto, A. Matsuzawa, and K. Murata, "A 10-b 20-MHz 30-mW pipelined interpolating CMOS ADC," *IEEE J. Solid-State Circuits*, vol. 28, pp. 1200–1206, 1993.

[19] K.Bult and A. Buchwald, "An embedded 240-mW 10-b 50 MS/s CMOS ADC in 1 mm^2," *IEEE J. Solid-State Circuits*, vol. 32, pp. 1887–1895, 1997.

[20] R. van de Grift, I. Rutten, and M. van der Veen, "An 8-bit video ADC incorporating folding and interpolation techniques," *IEEE J. Solid-State Circuits*, vol. 22, pp. 944–953, 1987.

[21] J. van Valburg and R. van de Plassche, "An 8-b 650-MHz folding ADC," *IEEE J. Solid-State Circuits*, vol. 27, pp. 1662–1666, 1992.

[22] B. Nauta and A. Venes, "A 70-MS/s 110-mW 8-b CMOS folding and interpolating A/D converter," *IEEE J. Solid-State Circuits*, vol. 30, pp. 1302–1308, 1995.

[23] P. Vorenkamp and R. Roovers, "A 12-b, 60-MSample/s cascaded folding and interpolating ADC," *IEEE J. Solid-State Circuits*, vol. 32, pp. 1876–1886, 1997.

[24] G. Hoogzaad and R. Roovers, "A 65-mW, 10-bit, 40-MSample/s BiCMOS Nyquist ADC in 0.8 mm^2," *IEEE J. Solid-State Circuits*, vol. 34, pp. 1796–1802, 1999.

[25] R. J. van de Plassche and R. E. J. van der Grift, "A high-speed 7 bit A/D converter," *IEEE J. Solid-State Circuits*, vol. 6, pp. 938–943, 1979.

[26] A. Dingwall and V. Zazzu, "An 8-MHz CMOS subranging 8-bit A/D converter," *IEEE J. Solid-State Circuits*, vol. 20, pp. 1138–1143, 1985.

[27] C. Moreland, F. Murden, M. Elliott, J. Young, M. Hensley, and R. Stop, "A 14-bit 100-MSample/s subranging ADC," *IEEE J. Solid-State Circuits*, vol. 35, pp. 1791–1798, 2000.

[28] H. van der Ploeg and R. Remmers, "A 3.3-V, 10-b, 25-MSample/s two-step ADC in 0.35-μm CMOS," *IEEE J. Solid-State Circuits*, vol. 34, pp. 1803–1811, 1999.

[29] S. Lewis and P. Gray, "A pipelined 5-MSample/s 9-bit analog-to-digital converter," *IEEE J. Solid-State Circuits*, vol. 22, pp. 954–961, 1987.

[30] I. Mehr and L. Singer, "A 55-mW, 10-bit, 40-MSample/s Nyquist-rate CMOS ADC," *IEEE J. Solid-State Circuits*, vol. 35, pp. 318–325, 2000.

[31] D. Cline and P. Gray, "A power optimized 13-b 5 MSample/s pipelined Analog-to-digital converter in 1.2 μm CMOS," *IEEE J. Solid-State Circuits*, vol. 31, pp. 294–303, 1996.

[32] J.-B. Shyu, G. Temes, and K. Yao, "Random errors in MOS capacitors," *IEEE J. Solid-State Circuits*, vol. 6, pp. 1070–1076, 1982.

[33] W. Yang, D. Kelly, I. Mehr, M. Sayuk, and L. Singer, "A 3-V 340-mW 14b 75-Msample/s CMOS ADC with85-dB SFDR at Nyquist input," *IEEE J. Solid-State Circuits*, vol. 36, pp. 1931–1936, 2001.

[34] A. Abo and P. Gray, "A 1.5-V, 10-bit, 14.3-MS/s CMOS pipeline anlog-to-digital converter," *IEEE J. Solid-State Circuits*, vol. 34, pp. 599–606, 1999.

[35] L. Singer, S. Ho, M. Timko, and D. Kelly, "A 12 b 65 MSample/s CMOS ADC with 82 dB SFDR at 120MHz," *ISSCC Digest of Technical Papers*, pp. 38–39, 2000.

[36] D. Kelly, I. Mehr, M. Sayuk, and L. Singer, "A 3V 340mW 14b 75MSps CMOS ADC with 85dB SFDR at Nyquist," *ISSCC Digest of Technical Papers*, pp. 134–135, 2001.

[37] B. Murmann and B. Boser, "A 12-bit 75MS/s pipelined ADC using open-loop residue amplification," *IEEE J. Solid-State Circuits*, vol. 38, pp. 2040–2050, 2003.

[38] J. McCreary and P. Gray, "All-MOS charge redistribution analog-to-digital conversion techniques-part I," *IEEE J. Solid-State Circuits*, vol. 10, pp. 371–379, 1975.

[39] K. Bacrania, "A 12-bit succesive-approximation-type ADC with digital error correction," *IEEE J. Solid-State Circuits*, vol. 21, pp. 1016–1025, 1986.

[40] F. Kuttner, "A 1.2 V 10 b 20 MSample/s non-binary succesive approximation ADC in 0.13μm CMOS," *ISSCC Digest of Technical Papers*, pp. 176–177, 2002.

[41] J. Lin and B. Haroun, "An embedded 0.8 V/480 μW 6b/22 MHz flash ADC in 0.13-μm digital CMOS process using a nonlinear double interpolation technique," *IEEE J. Solid-State Circuits*, vol. 37, pp. 1610–1617, 2002.

[42] K. Sushihara and A. Matsuzawa, "A 7 b 450 MSample/s 50 mW CMOS ADC in 0.3 mm^2," *ISSCC Digest of Technical Papers*, pp. 170–171, 2002.

[43] C. Donovan and M. Flynn, "A "digital" 6-bit ADC in 0.25-μm CMOS," *IEEE J. Solid-State Circuits*, vol. 37, pp. 432–437, 2002.

[44] I. Mehr and D. Dalton, "A 500-MSample/s, 6-bit Nyquist-rate ADC for disk-drive read-channel applications," *IEEE J. Solid-State Circuits*, vol. 34, pp. 912–920, 1999.

[45] K. Nagaraj, D. Martin, M. Wolfe, R. Chattopadhyay, S. Pavan, J. Cancio, and T. Viswanathan, "A dual-mode 700-MSample/s 6-bit 200-MSample/s 7-bit A/D converter in a 0.25-μm digital CMOS process," *IEEE J. Solid-State Circuits*, vol. 35, pp. 1760–1768, 2000.

[46] M. Choi and A. Abidi, "A 6 b 1.3 GSample/s A/D converter in 0.35 μm CMOS," *IEEE J. Solid-State Circuits*, vol. 36, pp. 1847–1858, 2001.

[47] K. Nagaraj, F. Chen, T. Le, and T. R. Viswanathan, "Efficient 6-bit A/D converter using a 1-bit folding front end," *IEEE J. Solid-State Circuits*, vol. 34, pp. 1056–1062, 1999.

[48] K. Uyttenhove, J. Vandenbussche, E. Lauwers, G. Gielen, and M. Steyaert, "Design techniques and implementation of an 8-bit 200-MS/s interpolating/averaging CMOS A/D converter," *IEEE J. Solid-State Circuits*, vol. 38, pp. 483–494, 2003.

[49] R. Taft, C. Menkus, M. Tursi, O. Hidri, and V. Pons, "A 1.8 V 1.6 GS/s 8b Self-Calibrating Folding ADC with 7.26 ENOB at Nyquist Frequency," *ISSCC Digest of Technical Papers*, vol. 47, pp. 252–253, 2004.

[50] G. Geelen and E. Paulus, "An 8 b 600 MS/s 200 mW CMOS Folding A/D Converter Using an Amplifier Preset Technique," *ISSCC Digest of Technical Papers*, vol. 47, pp. 254–255, 2004.

[51] K. Yoon, J. Lee, D.-K. Jeong, and W. Kim, "An 8-bit 125 MS/s CMOS folding ADC for gigabit ethernet LSI," *VLSI Digest of Technical Papers*, pp. 212–213, 2000.

[52] M.-H. Liu and S.-I. Liu, "An 8-bit 10 MS/s folding and interpolating ADC using the continous-time auto-zero technique," *IEEE J. Solid-State Circuits*, vol. 36, pp. 122–128, 2001.

[53] Y. Li and E. Sánchez-Sinencio, "A wide input bandwidth 7-bit 300-MSample/s folding and current-mode interpolating ADC," *IEEE J. Solid-State Circuits*, vol. 38, pp. 1405–1410, 2003.

[54] R. Taft and M. Tursi, "A 100-MS/s 8-b CMOS subranging ADC with sustained parametric performance from 3.8 V down to 2.2 V," *IEEE J. Solid-State Circuits*, vol. 36, pp. 331–338, 2001.

[55] B. Brandt and J. Lutsky, "A 75-mW. 10-b, 20-MSPS CMOS subranging ADC with 9.5 effective bits at Nyquist," *IEEE J. Solid-State Circuits*, vol. 34, pp. 1788–1795, 1999.

[56] H. Pan, M. Segami, M. Choi, J. Cao, and A. Abidi, "A 3.3-V 12-b 50-MS/s A/D converter in 0.6-μm CMOS with over 80-dB SFDR," *IEEE J. Solid-State Circuits*, vol. 35, pp. 1769–1780, 2000.

[57] M.-J. Choe, B.-S. Song, and K. Bacrania, "A 13-b 40-MSamples/s CMOS pipelined folding ADC with background offset trimming," *IEEE J. Solid-State Circuits*, vol. 35, pp. 1781–1790, 2000.

[58] H. van der Ploeg, G. Hoogzaad, H. Termeer, M. Vertregt, and R. Roovers, "A 2.5-V 12-b 54-MSample/s 0.25-μm CMOS ADC in 1 mm^2 with mixed-signal chopping and calibration," *IEEE J. Solid-State Circuits*, vol. 36, pp. 1859–1867, 2001.

[59] B. Hernes, A. Briskemyr, T. Andersen, F. Telstø, T. Bonnerud, and O. Moldsvor, "A 1.2V 220MS/s 10b Pipeline ADC Implemented in 0.13 m Digital CMOS," *ISSCC Digest of Technical Papers*, vol. 47, pp. 256–257, 2004.

[60] K. Poulton, R. Neff, A. Muto, W. Liu, A. Burstein, and M. Heshami, "A 4 GSample/s 8b ADC in 0.35 μm CMOS," *ISSCC Digest of Technical Papers*, pp. 166–167, 2002.

[61] K. Poulton, R. Neff, B. Setterberg, B. Wuppermann, T. Kopley, R. Jewett, J. Pernillo, C. Tan, and A. Montijo, "A 20 GS/s 8 b ADC with a 1 MB memory in 0.18 μm CMOS," *ISSCC Digest of Technical Papers*, pp. 318–319, 2003.

[62] S.-M. Yoo, J.-B. Park, H.-S. Yang, H.-H. Bae, K.-H. Moon, H.-J. Park, S.-H. Lee, and J.-H. Kim, "A 10 b 150 MS/s 123 mW 0.18 μm CMOS pipelined ADC," *ISSCC Digest of Technical Papers*, pp. 326–327, 2003.

[63] K. Kaviani, O. Oralkan, P. Khuri-Yakub, and B. Wooley, "A multichannel pipeline analog-to-digital converter for an integrated 3-D ultrasound imaging system," *IEEE J. Solid-State Circuits*, vol. 38, pp. 1266–1270, 2003.

[64] D. Miyazaki, S. Kawahito, and M. Furuta, "A 10-b 30-MS/s low-power pipelined CMOS A/D converter using a pseudodifferential architecture," *IEEE J. Solid-State Circuits*, vol. 38, pp. 369–373, 2003.

[65] Y.-I. Park, S. Karthikeyan, F. Tsay, and E. Bartolome, "A 10 b 100 MSample/s CMOS pipelined ADC with 1.8 V power supply," *ISSCC Digest of Technical Papers*, pp. 130–131, 2001.

[66] D. Miyazaki, M. Furuta, and S. Kawahito, "A 16 mW 30 MSample/s 10 b pipelined A/D converter using a pseudo-differential architecture," *ISSCC Digest of Technical Papers*, pp. 174–175, 2002.

[67] B.-M. Min, P. Kim, F. Bowman, D. Boisvert, and A. Aude, "A 69-mW 10-bit 80-MSample/s pipelined CMOS ADC," *IEEE J. Solid-State Circuits*, vol. 38, pp. 2031–2039, 2003.

[68] K. Irie, N. Kusayanagi, T. Kawachi, T. Nishibu, and Y. Matsumori, "An 8 b 500 MS/s full Nyquist cascade A/D converter," *VLSI Digest of Technical Papers*, pp. 77–78, 1999.

[69] H.-C. Liu, Z.-M. Lee, and J.-T. Wu, "A 15 b 20 MS/s CMOS Pipelined ADC with Digital Background Calibration," *ISSCC Digest of Technical Papers*, vol. 47, pp. 454–455, 2004.

[70] I. Opris, L. Lewicki, and B. Wong, "A single-ended 12-bit 20 MSample/s self-calibrating pipeline A/D converter," *IEEE J. Solid-State Circuits*, vol. 33, pp. 1898–1903, 1998.

[71] J. Ingino and B. Wooley, "A continuously calibrated 12-b 10-MS/s, 3.3-V A/D converter," *IEEE J. Solid-State Circuits*, vol. 33, pp. 1920–1931, 1998.

[72] A. Shabra and H.-S. Lee, "Oversampled pipeline A/D converters with mismatch shaping," *IEEE J. Solid-State Circuits*, vol. 37, pp. 566–578, 2002.

[73] H.-S. Chen, K. Bacrania, and B.-S. Song, "A 14 b 20 MSample/s CMOS pipelined ADC," *ISSCC Digest of Technical Papers*, pp. 46–47, 2000.

[74] I. Opris, B. Wong, and S. Chin, "A pipeline A/D converter architecture with low DNL," *IEEE J. Solid-State Circuits*, vol. 35, pp. 281–285, 2000.

[75] K. Nair and R. Harjani, "A 96 dB SFDR 50 MS/s Digitally Enhanced CMOS Pipeline A/D Converter," *ISSCC Digest of Technical Papers*, vol. 47, pp. 456–457, 2004.

[76] S. Limotyrakis, S. Kulchycki, D. Su, and B. Wooley, "A 150 MS/s 8 b 71 mW Time-Interleaved ADC in 0.18 m CMOS," *ISSCC Digest of Technical Papers*, vol. 47, pp. 258–259, 2004.

[77] Y. Chiu, P. Gray, and B. Nikolic, "A 1.8 V 14 b 10 MS/s Pipelined ADC in 0.18 m CMOS with 99 dB SFDR," *ISSCC Digest of Technical Papers*, vol. 47, pp. 458–459, 2004.

[78] E. Siragusa and I. Galton, "A Digitally Enhanced 1.8 V 15 b 40 MS/s CMOS Pipelined ADC," *ISSCC Digest of Technical Papers*, vol. 47, pp. 452–453, 2004.

[79] C. Michalski, "A 12 b 105 MSample/s, 850 mW analog to digital converter," *VLSI Digest of Technical Papers*, pp. 208–211, 2000.

[80] J. Sauerbrey, D. Schmitt-Landsiedel, and R. Thewes, "A 0.5-V 1-μW successive approximation ADC," *IEEE J. Solid-State Circuits*, vol. 38, pp. 1261–1265, 2003.

[81] M. Scott, B. Boser, and K. Pister, "An ultralow-energy ADC for smart dust," *IEEE J. Solid-State Circuits*, vol. 38, pp. 1123–1129, 2003.

[82] E. Blecker, T. McDonald, O. Erdoğan, P. Hurst, and S. Lewis, "Digital background calibration of an algorithmic analog-to-digital converter using a simplified queue," *IEEE J. Solid-State Circuits*, vol. 38, pp. 1059–1062, 2003.

[83] P. Rombouts, W. de Bilde, and L. Weyten, "A 13.5-b 1.2-V micropower extended counting A/D converter," *IEEE J. Solid-State Circuits*, vol. 36, pp. 176–183, 2001.

[84] O. Erdoğan, P. Hurst, and S. Lewis, "A 12-b digital-background-calibrated algorithmic ADC with -90-dB THD," *IEEE J. Solid-State Circuits*, vol. 34, pp. 1812–1820, 1999.

[85] M. Pelgrom, "A 10-b 50-MHz CMOS D/A Converter with 75Ω Buffer," *IEEE J. Solid-State Circuits*, vol. 25, pp. 1347–1352, 19910.

[86] H. P. Tuinhout, G. Hoogzaad, M. Vertregt, R. L. J. Roovers, and C. Erdmann, "Design and characterization of a high-precision resistor ladder test structure," *IEEE Transactions on Semiconductor Manufacturing*, vol. 16, pp. 187–193, 2003.

[87] B. Razavi, *Principles of data conversion system design*. Piscataway: IEEE Press, 1995.

[88] T. Shimizu, M. Hotta, K. Maio, and S. Ueda, "A 10-bit 20-MHz two-step parallel A/D converter with internal S/H," *IEEE J. Solid-State Circuits*, vol. 24, pp. 13–20, 1989.

[89] P. Vorenkamp and J. Verdaasdonk, "A 10 b 50 MS/s pipelined ADC," *ISSCC Digest of Technical Papers*, pp. 32–33, 1992.

[90] W. Colleran and A. Abidi, "A 10-b, 75-MHz two-stage pipelined bipolar A/D converter," *IEEE J. Solid-State Circuits*, vol. 28, pp. 1187–1199, 1993.

[91] P. Grant and K. Smith, "Monotonic dual-ladder A/D conversion," *IEEE J. Solid-State Circuits*, vol. 22, pp. 295–297, 1987.

[92] M. Kolluri, "A 12-bit 500-ns subranging ADC," *IEEE J. Solid-State Circuits*, vol. 24, pp. 1498–1506, 1989.

[93] G. Erdi, "A percision trim technique for monolithic analog circuits," *IEEE J. Solid-State Circuits*, vol. 10, pp. 412–416, 1975.

[94] M. Mayes, S. Chin, and L. Stoian, "A low-power 1 MHz. 25 mW 12-bit time-interleaved analog-to-digital converter," *IEEE J. Solid-State Circuits*, vol. 31, pp. 169–178, 1996.

[95] K. Kim, N. Kusayanagi, and A. Abidi, "A 10-b, 100-MS/s CMOS A/D converter," *IEEE J. Solid-State Circuits*, vol. 32, pp. 302–311, 1997.

[96] D. Mercer, "A 14-b, 2.5 MSps pipelined ADC with on-chip EPROM," *IEEE J. Solid-State Circuits*, vol. 31, pp. 70–76, 1996.

[97] H.-S. Lee, D. Hodges, and P. Gray, "A self-calibrating 15 bit CMOS A/D converter," *IEEE J. Solid-State Circuits*, vol. 19, pp. 813–819, 1984.

[98] S.-H. Lee and B.-S. Song, "Digital-domain calibration of multistep analog-to-digital converters," *IEEE J. Solid-State Circuits*, vol. 27, pp. 1679–1688, 1992.

[99] A. Karanicolas, H.-S. Lee, and K. Bacrania, "A 15-b 1-MSample/s digitally self-calibrated pipeline ADC," *IEEE J. Solid-State Circuits*, vol. 28, pp. 1207–1215, 1993.

[100] H.-S. Lee, "A 12-b 600 k/s digitally self-calibrated pipelined algorithmic ADC," *IEEE J. Solid-State Circuits*, vol. 29, pp. 509–515, 1994.

[101] T.-H. Shu, B.-S. Song, and K. Bacrania, "A 13-b 10-MSample/s ADC digitally calibrated with oversampling delta-sigma converter," *IEEE J. Solid-State Circuits*, vol. 30, pp. 443–452, 1995.

[102] S.-Y. Chuang and T. Sculley, "A digitally self-calibrating 14-bit 10-MHz CMOS pipelined A/D converter," *IEEE J. Solid-State Circuits*, vol. 37, pp. 674–683, 2002.

[103] Y.-M. Lin, B. Kim, and P. Gray, "A 13-b 2.5-MHz self-calibrated pipelined A/D convertre in 3-μm CMOS," *IEEE J. Solid-State Circuits*, vol. 26, pp. 628–636, 1991.

[104] B. Razavi and B. Wooley, "Design techniques for high-speed high-resolution comparators," *IEEE J. Solid-State Circuits*, vol. 27, pp. 1916–1926, 1992.

[105] B.-S. Song, M. Tompsett, and K. Lakshmikumar, "A 12-bit 1-MSample/s capacitor error-averaging pipelined A/D converter," *IEEE J. Solid-State Circuits*, vol. 23, pp. 1324–1333, 1988.

[106] E. Sackinger and W. Guggenbuhl, "A versatile building block: the CMOS differential difference amplifier," *IEEE J. Solid-State Circuits*, vol. 22, pp. 287–294, 1987.

[107] K. Dyer, D. Fu, S. Lewis, and P. Hurst, "An analog background calibration technique for time interleaved analog to digital converters," *IEEE J. Solid-State Circuits*, vol. 33, pp. 1912–1919, 1998.

[108] K. Yoon, S. Park, and W. Kim, "A 6 b 500 MSample/s CMOS flash ADC with a background interpolated auto-zeroing technique," *ISSCC Digest of Technical Papers*, pp. 326–327, 1999.

[109] S.-U. Kwak, H.-S. Lee, and K. Bacrania, "A 15-b, 5-MSample/s low-spurious CMOS ADC," *IEEE J. Solid-State Circuits*, vol. 32, pp. 1866–1875, 1997.

[110] P. Yu and H.-S. Lee, "A 2.5-V, 12-b, 5-MSample/s pipelined CMOS ADC," *IEEE J. Solid-State Circuits*, vol. 31, pp. 1854–1861, 1996.

[111] P. Yu, S. Shehata, A. Joharapurkar, P. Chugh, A. Bugeja, X. Du, S.-U. Kwak, Y. Panantonopoulous, and T. Kuyel, "A 14b 40MSample/s pipelined ADC with DFCA," *ISSCC Digest of Technical Papers*, pp. 136–137, 2001.

[112] D. Fu, K. Dyer, S. Lewis, and P. Hurst, "A digital background calibration technique for time interleaved analog to digital converters," *IEEE J. Solid-State Circuits*, vol. 33, pp. 1904–1911, 1998.

[113] I. Galton, "Digital Cancellation of D/A Converter Noise in Pipelined A/D Converters," *IEEE Transaction on Circuits and Systems*, vol. 47, pp. 185–196, 2000.

[114] J. MacWilliams and N. Sloane, "Pseudo-Random Sequences and Arrays," *Proceedings of the IEEE*, vol. 64, p. 1715, 1976.

[115] S.M. Yamal, D. Fu, N. C.-J. Chang, P. Hurst, and S. Lewis, "Oversampled pipeline A/D converters with mismatch shaping," *IEEE J. Solid-State Circuits*, vol. 37, pp. 566–578, 2002.

[116] C.P. Sandbank, *Digital television*. Chichester: John Wiley & Sons Ltd., 1990.

[117] J. Doernberg, P. Gray, and D. Hodges, "A 10-bit 5-Msample/s CMOS two-step flash ADC," *IEEE J. Solid-State Circuits*, vol. 24, pp. 241–249, 1989.

[118] K. Stafford, P. Gary, and R. Blanchard, "A complete monolithic sample/hold amplifier," *IEEE J. Solid-State Circuits*, vol. 9, pp. 381–387, 1974.

[119] W. Liu, K.-B. Thei, H.-M. Chuang, K.-W. Lin, C.-C. Cheng, Y.-S. Ho, C.-W. Su, S.-C. Wong, C.-H. Lin, and C. Diaz, "Characterization of Polysilicon Resistors in Sub-0.25μm CMOS ULSI Applications," *IEEE Electron Device Letters*, vol. 22, pp. 318–320, 2001.

[120] J. Rijns, "CMOS low distortion high-frequency variable-gain amplifier," *IEEE J. Solid-State Circuits*, vol. 31, pp. 1029–1035, 1996.

[121] T. Miki, H. Kouno, Y. Kinoshita, T. Igarashi, and K. Okada, "A 10-b 50 MS/s 500-mW A/D converter using a differential-voltage subconverter," *IEEE J. Solid-State Circuits*, vol. 29, pp. 516–522, 1994.

[122] W. Black and D. Hodges, "Time interleaved converter arrays," *IEEE J. Solid-State Circuits*, vol. 15, pp. 1022–1029, 1980.

[123] A. Zjajo, H. van der Ploeg, and M. Vertregt, "A 1.8V 100mW 12-bits 80Msample/s two-step ADC in 0.18-μm CMOS," *Proceedings of ESSIRC*, pp. 241–244, 2003.

[124] R. J. van de Plassche and P. Baltus, "An 8-bit 100-MHz full-Nyquist analog-to-digital converter," *IEEE J. Solid-State Circuits*, vol. 23, pp. 1334–1344, 1988.

[125] T. Miki, Y. Nakamura, M. Nakaya, S. Asai, Y. Akasaka, and Y. Horiba, "An 80-MHz 8-bit CMOS D/A Converter," *IEEE J. Solid-State Circuits*, vol. 21, pp. 983–988, 1986.

[126] S.-B. You, J.-W. Kim, and S. Kim, "A CMOS 16-bit 20MSPS analog front end for scanner/MFP applications," *IEEE Transactions on Consumer Electronics*, vol. 49, pp. 647–652, 2003.

Index